Grade 4

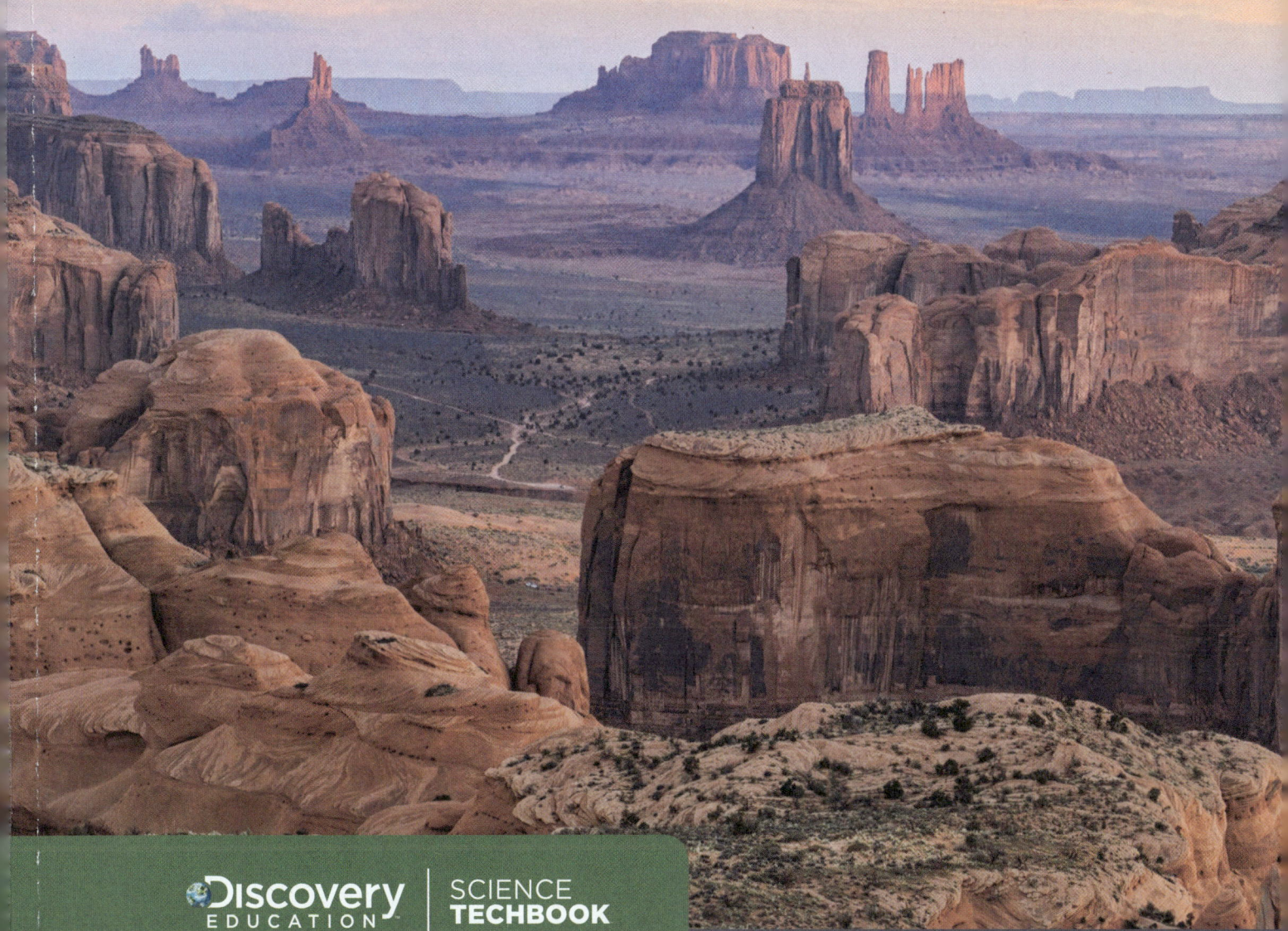

Discovery Education | SCIENCE TECHBOOK

Unit 3
Earth's Changing Surface

Discovery EDUCATION

Copyright © 2020 by Discovery Education, Inc. All rights reserved. No part of this work may be reproduced, distributed, or transmitted in any form or by any means, or stored in a retrieval or database system, without the prior written permission of Discovery Education, Inc.

NGSS is a registered trademark of Achieve. Neither Achieve nor the lead states and partners that developed the Next Generation Science Standards were involved in the production of this product, and do not endorse it.

To obtain permission(s) or for inquiries, submit a request to:

Discovery Education, Inc.
4350 Congress Street, Suite 700
Charlotte, NC 28209
800-323-9084
Education_Info@DiscoveryEd.com

ISBN 13: 978-1-68220-800-7

Printed in the United States of America.

6 7 8 9 10 CWM 26 25 24 23   B

© Discovery Education | www.discoveryeducation.com

Acknowledgments

Acknowledgment is given to photographers, artists, and agents for permission to feature their copyrighted material.

Cover and inside cover art: Zhukova Valentyna / Shutterstock.com

Table of Contents

Unit 3: Earth's Changing Surface

Letter to the Parent/Guardian vi

Unit Overview 1

    Anchor Phenomenon: The Grand Canyon from Space 2

Unit Project Preview: Lava Flows and the Grand Canyon 4

Concept 3.1 Breaking Down and Moving Rocks

Concept Overview 6

    Wonder 8

        Investigative Phenomenon: Disappearing Sandcastles 10

    Learn 16

    Share 38

Concept 3.2 Changing Landscapes

Concept Overview 48

    Wonder 50

        Investigative Phenomenon: Canyons 52

    Learn 64

    Share 96

Concept 3.3 Mapping Landforms

Concept Overview 106

   Wonder 108

      Investigative Phenomenon: Satellite Image of an Ocean 110

   Learn 118

   Share 140

Concept 3.4 Volcanoes

Concept Overview 150

   Wonder 152

      Investigative Phenomenon: Landscape Safety Check 154

   Learn 162

   Share 186

Unit Wrap-Up

Unit Project: Lava Flows and the Grand Canyon 196

Grade 4 Resources

Bubble Map R3

Safety in the Science Classroom R4

Vocabulary Flash Cards R7

Glossary R19

Index R50

Unit 3: Earth's Changing Surface

Dear Parent/Guardian,

This year, your student will be using Science Techbook™, a comprehensive science program developed by the educators and designers at Discovery Education and written to the Next Generation Science Standards (NGSS). The NGSS expect students to act and think like scientists and engineers, to ask questions about the world around them, and to solve real-world problems through the application of critical thinking across the domains of science (Life Science, Earth and Space Science, Physical Science).

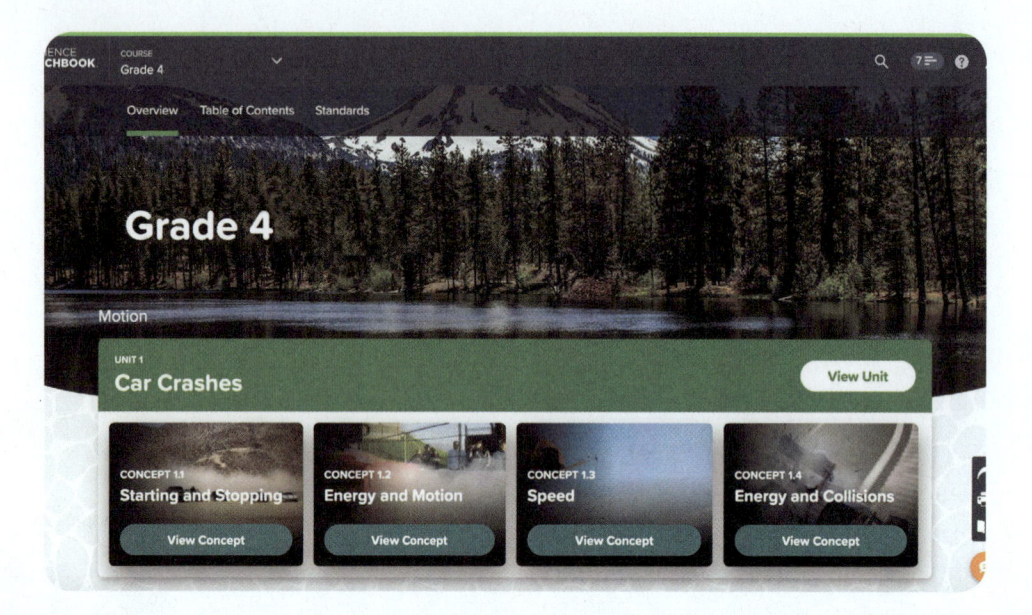

Science Techbook is an innovative program that helps your student master key scientific concepts. Students engage with interactive science materials to analyze and interpret data, think critically, solve problems, and make connections across science disciplines. Science Techbook includes dynamic content, videos, digital tools, Hands-On Activities and labs, and game-like activities that inspire and motivate scientific learning and curiosity.

You and your child can access the resource by signing in to www.discoveryeducation.com. You can view your child's progress in the course by selecting the Assignment button.

Science Techbook is divided into units, and each unit is divided into concepts. Each concept has three sections: Wonder, Learn, and Share.

Units and Concepts Students begin to consider the connections across fields of science to understand, analyze, and describe real-world phenomena.

Wonder Students activate their prior knowledge of a concept's essential ideas and begin making connections to a real-world phenomenon and the **Can You Explain?** question.

Learn Students dive deeper into how real-world science phenomenon works through critical reading of the Core Interactive Text. Students also build their learning through Hands-On Activities and interactives focused on the learning goals.

Share Students share their learning with their teacher and classmates using evidence they have gathered and analyzed during Learn. Students connect their learning with STEM careers and problem-solving skills.

Within this Student Edition, you'll find QR codes and quick codes that take you and your student to a corresponding section of Science Techbook online. To use the QR codes, you'll need to download a free QR reader. Readers are available for phones, tablets, laptops, desktops, and other devices. Most use the device's camera, but there are some that scan documents that are on your screen.

For resources in Science Techbook, you'll need to sign in with your student's username and password the first time you access a QR code. After that, you won't need to sign in again, unless you log out or remain inactive for too long.

We encourage you to support your student in using the print and online interactive materials in Science Techbook, on any device. Together, may you and your student enjoy a fantastic year of science!

Sincerely,

The Discovery Education Science Team

Unit 3: Earth's Changing Surface | vii

© Discovery Education | www.discoveryeducation.com • Image: Zhukova Valentyna / Shutterstock.com

Unit 3
Earth's Changing Surface

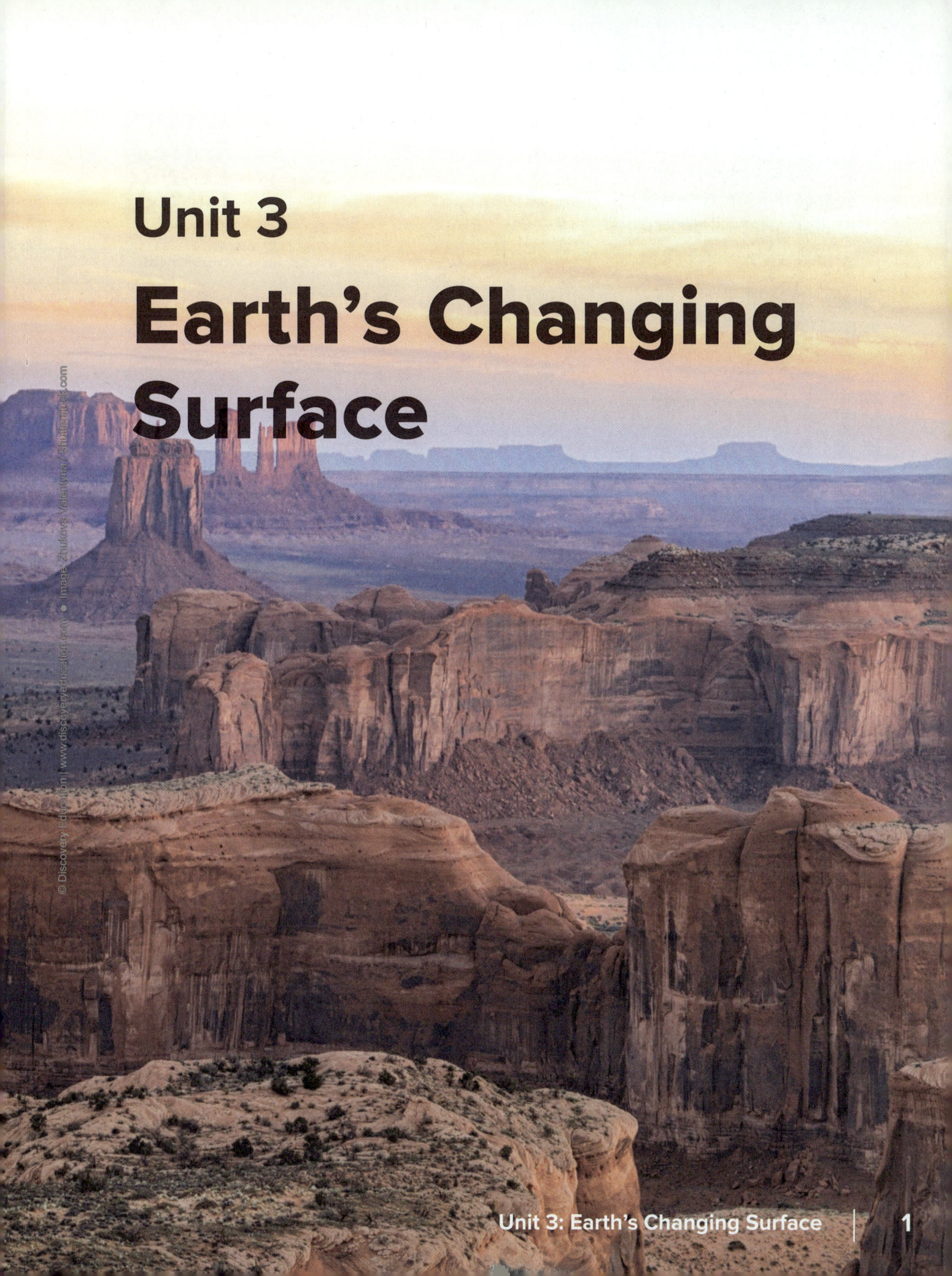

Get Started

The Grand Canyon from Space

The Grand Canyon appears from space to be a giant gash in the surface of Earth. How was this immense canyon formed? Why does it have so many different colors?

Quick Code: us4505s

The Grand Canyon from Space

Think About It

Look at the photograph. **Think** about the following questions.

- How do water, ice, wind, and vegetation sculpt landscapes?
- What factors affect how quickly landscapes change?
- How are landscape changes recorded by layers of rocks and fossils?
- How can people protect themselves and the environment from the impact of changing landscapes?

The Grand Canyon

Unit 3: Earth's Changing Surface | 3

Unit Project Preview

Solve Problems Like a Scientist

Unit Project: Lava Flows and the Grand Canyon

In this project, you will use what you know about Earth's surface changes to model how lava affected the Grand Canyon.

Quick Code: us4506s

Grand Canyon Lava Flow

SEP Developing and Using Models
CCC Cause and Effect

Ask Questions About the Problem

You are going to build a model to predict what will happen if lava flows into the Grand Canyon. **Write** some questions you can ask to learn more about the problem. As you learn about landforms and the forces that shape them in this unit, **write** down the answers to your questions.

Unit 3: Earth's Changing Surface | 5

CONCEPT 3.1

Breaking Down and Moving Rocks

Student Objectives

By the end of this lesson:

- ☐ I can explain the roles water, wind, and heat play in weathering, erosion, and deposition.
- ☐ I can provide evidence that mechanical and chemical weathering change Earth's surface over time.

Key Vocabulary

- ☐ air
- ☐ chemical weathering
- ☐ deposition
- ☐ erosion
- ☐ heat
- ☐ sediment
- ☐ soil
- ☐ water
- ☐ weathering

Quick Code: us4508s

Concept 3.1: Breaking Down and Moving Rocks

Activity 1
Can You Explain?

How do wind, water, and weather erode Earth's surface?

Quick Code: us4509s

Concept 3.1: Breaking Down and Moving Rocks

3.1 | Wonder
How do wind, water, and weather erode Earth's surface?

Activity 2
Ask Questions Like a Scientist

Disappearing Sandcastles

Quick Code: us4510s

Look at the images of a sandcastle and a beach. **Think** about these questions: Have you ever made a sandcastle? What happened to it?

Let's Investigate Disappearing Sandcastles

Beach Erosion

SEP Asking Questions and Defining Problems
CCC Cause and Effect

10

What do you wonder about these images? **Think** about how different agents break rocks and sediments and move them around. **Write** three questions you have and **share** them with the class.

I wonder... *why do I not fell the rock in the sens*

I wonder... *why the sens Bronwen but it tousted tellow*

I wonder... *why the send have trass in it.*

Concept 3.1: Breaking Down and Moving Rocks

3.1 | Wonder How do wind, water, and weather erode Earth's surface?

Activity 3
Observe Like a Scientist

Sandcastles, Rocks, and Canyons

Quick Code: us4511s

Observe the images. **Look** for the various shapes. Then, **answer** the questions that follow.

Wrecked Sandcastle

Coastal Rocks

Bryce Canyon

CCC Patterns

Discuss with Your Class

Look at the image Wrecked Sandcastle. Are there any parts of the castle that remind you of other landscapes that you have seen before?

_____
_____
_____
_____
_____
_____
_____

Compare the images Wrecked Sandcastle and Coastal Rocks. Can you see any similarities between the two images? Why do you think the rock is shaped this way? What forces do you think created these shapes?

_____
_____
_____
_____
_____
_____
_____

Concept 3.1: Breaking Down and Moving Rocks

3.1 | Wonder   How do wind, water, and weather erode Earth's surface?

Look at the image Bryce Canyon. How do you think the canyon was formed? Can you think of a reason why the surrounding rocks are rounded and not jagged?

_____

_____

_____

_____

_____

_____

What did each scene look like 20 minutes before the picture was taken?
What did each scene look like one hour after the picture was taken?
What would each scene look like 10 years from now?

_____

_____

_____

_____

_____

_____

Activity 4
Evaluate Like a Scientist

Quick Code: us4512s

What Do You Already Know About Breaking Down and Moving Rocks?

Shaping the Earth

Write each term from the left column in the correct place on the image. Then, **write** each definition from the right column under the correct term.

Deposition	Breaking apart rocks
Erosion	Laying sediment down
Weathering	Moving rock or soil around

Then, **write** each of these definitions next to the correct term, on the image:

Concept 3.1: Breaking Down and Moving Rocks   15

3.1 | Learn How do wind, water, and weather erode Earth's surface?

How Are Rocks Broken Down?

Activity 5

Observe Like a Scientist

What Is Weathering?

Watch the video to learn more about weathering. **Look** for landforms and objects that have been weathered.

Quick Code: us4513s

What Is Weathering? (Video)

Talk Together

Now, talk together about how weathering has changed the landforms or objects in the video. How is weathering different from weather? What does weathering have to do with the formation of landscapes?

Activity 6
Analyze Like a Scientist

Types of Weathering

Read the text and **look** at the image about weathering. As you read, **circle** the causes of each type of weathering and **underline** the effects of each.

Quick Code: us4514s

mountain
↓
boulders
↓
Rocks
↓
sand

Types of Weathering

mountain

The surface of Earth is always changing. One way that landforms change is by **weathering**. Weathering is what happens when rocks break apart. The enormous rocks that make up mountains can break into boulders. These boulders can further break down into smaller rocks, and the smaller rocks can continue to break apart to form sand. You have seen rocks of all different sizes—this is evidence of weathering.

Weathering has many causes. One of these causes is **water**. As water runs over rocks, it can dissolve some of the substances in them. This makes the rocks fall apart. Sometimes it completely dissolves the rock. Most caves are created by this type of weathering. **Air** can also change rocks, causing them to crumble. These are examples of **chemical weathering**.

Heat and cold also cause rocks to break. Water and cold temperatures often work together. Water can seep into tiny cracks in rocks and freeze when the temperature gets very cold. When water freezes, it expands. This makes the cracks even wider. If this process happens often enough, it can cause the rock to break apart.

CCC Cause and Effect

Concept 3.1: Breaking Down and Moving Rocks | 17

heat/cold and trees

Types of Weathering *cont'd*

Did you know that trees can break rocks, too? The roots of trees and other plants often grow into the cracks in a rock. As the roots continue to grow, they can eventually break the rock into pieces. These types of weathering are known as mechanical weathering. Since weathering happens over long periods of time, it is hard to see it in action. But you can see the effects of it all around you.

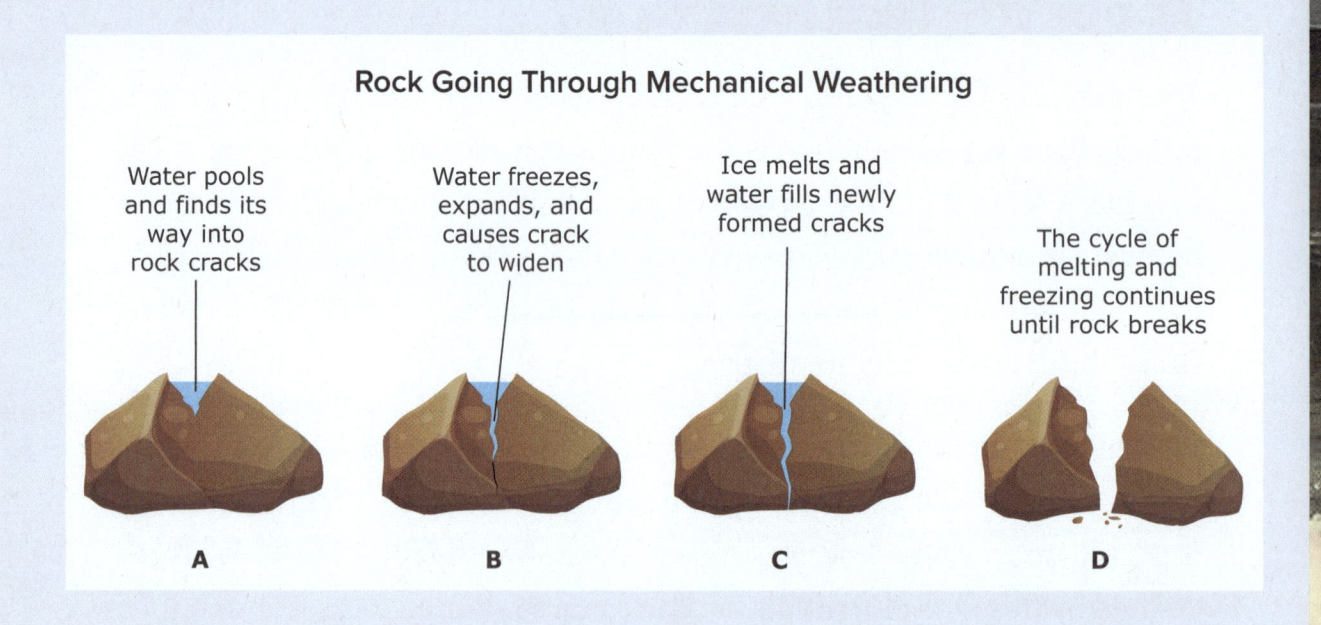

Rock Going Through Mechanical Weathering

A: Water pools and finds its way into rock cracks

B: Water freezes, expands, and causes crack to widen

C: Ice melts and water fills newly formed cracks

D: The cycle of melting and freezing continues until rock breaks

Now, **write** the causes and effects of chemical and mechanical weathering in the T-Chart.

Chemical Weathering	Mechanical Weathering

Concept 3.1: Breaking Down and Moving Rocks | 19

3.1 | Learn How do wind, water, and weather erode Earth's surface?

Activity 7
Observe Like a Scientist

Forces That Shape Earth

Quick Code: us4515s

Write your answers to the questions in the following chart, in the My Prediction column. Use what you already know.

	My Prediction	Is My Prediction Correct?
Which agent can change both sand and rock formations over time?		
Which agents can cause cracks in rock formations over time?		
Which is the slowest agent of mechanical weathering?		
Which agent can actually create an opening in a rock formation over time?		
What observations did you make as to how ice deposited sediments over time?		

Now, **complete** the Interactive to learn about the different causes and effects of mechanical weathering. **Look** for evidence that does or does not support your predictions. Based on the evidence, write Yes or No in the Is My Prediction Correct? column.

Forces That Shape Earth

Concept 3.1: Breaking Down and Moving Rocks

3.1 | Learn How do wind, water, and weather erode Earth's surface?

Activity 8
Investigate Like a Scientist

Quick Code: us4516s

Hands-On Investigation: Modeling Mechanical and Chemical Weathering

In this investigation, you will model and explore mechanical and chemical weathering to see the similarities and differences between the two.

Make a Prediction

First, **write** your predictions in the chart.

Question	My Prediction
Which means of weathering produced greater changes? Explain.	
How were the chemical weathering and mechanical weathering similar?	
How might the data you collected in the lab be useful in real-world applications?	

22

What materials do you need? (per group)

- Crackers (per student), 2
- Plastic cup, 9 oz
- Antacid tablets
- Writing utensil (per student)
- Napkin (per student)

What Will You Do?

1. Choose how you will model mechanical weathering.
2. Use one of the crackers to model mechanical weathering.
3. Record the results in your notebook.
4. Clean up the cracker crumbs.
5. Choose how you will model chemical weathering.
6. Then, use the other cracker to model chemical weathering.
7. Record the results in the notebook.
8. Clean up the cracker paste.
9. Finally, write a word on the board to describe chemical or mechanical weathering.

SEP Developing and Using Models

3.1 | Learn — How do wind, water, and weather erode Earth's surface?

Think About the Activity

Now, **examine** your model. **Think** about some critiques and **write** them in the T-Chart, together with possible ways to improve it.

Critiques	Suggested Improvements

Activity 9
Observe Like a Scientist

Quick Code: us4517s

Chemical and Mechanical Weathering

Watch the videos to learn more about the different types of weathering. **Look** for examples of chemical and mechanical weathering. Continue to **record** evidence about mechanical and chemical weathering in your T-Chart from Activity 6. Then, **answer** the questions.

Video
Mechanical Weathering

Video
Chemical Weathering

SEP Planning and Carrying Out Investigations
CCC Stability and Change

Concept 3.1: Breaking Down and Moving Rocks | 25

3.1 | Learn   How do wind, water, and weather erode Earth's surface?

Do you think chemical weathering or mechanical weathering happens faster?

_____
_____
_____
_____
_____
_____
_____

How might you find and collect data to see if you are correct?

_____
_____
_____
_____
_____
_____
_____

Activity 10
Observe Like a Scientist

Badlands Weathering

Observe the photograph of the Badlands in South Dakota. Then, **complete** the chart that follows. Use evidence from the activities you have done to support your claim.

Quick Code: us4518s

Badlands Weathering

SEP Analyzing and Interpreting Data

Concept 3.1: Breaking Down and Moving Rocks | 27

3.1 | Learn
How do wind, water, and weather erode Earth's surface?

Is the landform a result of mechanical or chemical weathering?

My claim (the answer to the question)

Evidence I found: Record all the evidence you gathered from video, reading, interactives, and hands-on investigations.

My claim is true because:

What Is Erosion, and How Does It Happen?

Activity 11
Analyze Like a Scientist

Quick Code: us4519s

Erosion

Read the text about erosion and **draw** an illustration for each paragraph. **Watch** the video for inspiration.

Erosion

After rocks are weathered, they can erode. **Erosion** is the process that occurs when sand, soil, or rocks are moved from one place to another. Gravity pulls rocks down mountainsides. Rivers erode rocks and **soil** from their banks and carry them downstream. Waves pull sand away from beaches. Little by little, rain washes the soil on hilly farmland downhill. Glaciers pick up and carry rocks, soil, and even large boulders in slow-moving rivers of ice and snow. The pieces of weathered rock that are moved by gravity, wind, water, and glaciers are called **sediments**.

> Erosion
> movement from one place to another caused by gravity or water

Concept 3.1: Breaking Down and Moving Rocks

Sometimes you can see erosion happening, such as during flash floods, hurricanes, or landslides. You may see sediments carried down gutters by water runoff after a big rainstorm. Or perhaps you have seen that sometimes the water in a nearby creek appears muddy. But most of the time you can see only the evidence left behind by hundreds, thousands, or millions of years of erosion. Grains of sand are blown by the wind a few feet at a time. Glaciers move rocks inch by inch. Over time, these little changes add up!

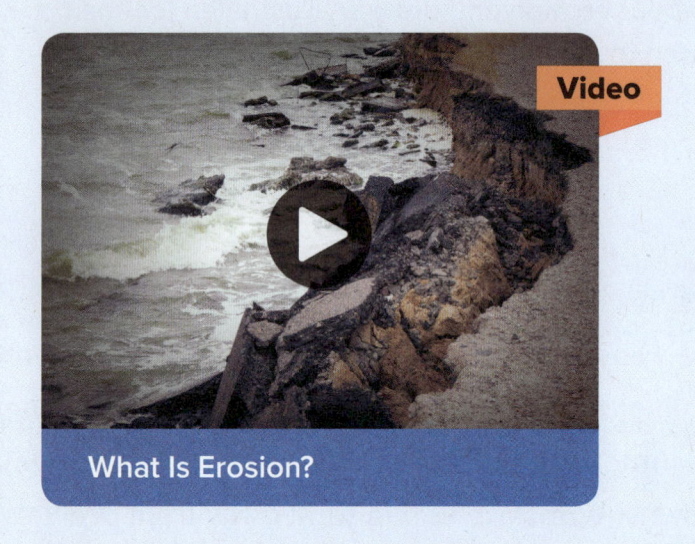

Video

What Is Erosion?

Activity 12

Investigate Like a Scientist

Quick Code: us4520s

Hands-On Investigation: Glacier Erosion

In this investigation, you will work with a group to create a model showing the effects of glacial erosion. Then, you will draw a picture of the model indicating how the glacier causes erosion. First, **write** your predictions in the chart.

Make a Prediction

Question	My Prediction
Imagine you were searching for a site where a glacier once flowed. What evidence might you observe that would indicate a glacier was once present?	
What happens to Earth's surface as a glacier moves across?	

SEP Developing and Using Models

Concept 3.1: Breaking Down and Moving Rocks

3.1 | Learn How do wind, water, and weather erode Earth's surface?

What materials do you need? (per group)

- Ice cubes, 3 to 4
- Sand
- Modeling clay
- Paper towels
- Aluminum foil pan, 13 × 9 × 2

What Will You Do?

As you **complete** the following steps, **write** your observations in your notebook.

1. Press an ice cube on the flat surface of the modeling clay. Move it back and forth several times. Does anything happen to the clay? To the ice?

2. Place a small pile of sand on the surface of the clay. Place the ice cube over the sand on the clay. Let it sit for about one minute. Pick up the ice cube and look at the surface that had been on the sand. Describe what you see. Hypothesize what will happen when this piece of ice is rubbed over the surface of the clay.

3. Now, place the ice cube back in the same position on the sandy surface of the clay and move the ice back and forth a few times. Remove the ice cube and gently wipe the excess sand off the surface of the clay. Describe the surface of the clay where it was rubbed by the sand and ice. Was your hypothesis correct or incorrect?

4. Then, **draw** a picture of the model. Show how the glacier causes erosion.

Think About the Activity

Now, **examine** your model. Identify its limitations and **write** them in the T-Chart. Then, **write** possible solutions to the limitations.

Limitations	Solutions

Concept 3.1: Breaking Down and Moving Rocks

What Happens to Rock Once It Is Eroded?

Activity 13
Analyze Like a Scientist

Deposition

Quick Code: us4521s

Read the text about deposition three times. The first time, **discuss** with your partner what it reminds you of. The second time, **underline** the main idea of the text. After the third time, discuss with your partner the passage in the text that says: "Erosion and deposition are connected." Use the chart to **explain** the cause-and-effect relationship between erosion and deposition.

Deposition

Erosion moves rocks and soil around, but deposition is the process that lays them back down. At some point, the wind, ice, or water will deposit the sediment it is carrying somewhere else. It will fall onto land that is already there or settle out on the bottom of a lake or the sea. If you see a deposit of sand, then you know it has already been eroded somewhere else. If rocks become eroded, then eventually they must be deposited. Erosion and deposition are connected. Sediments are the remains of weathered and eroded rock that have been deposited. Sediments build new landforms. A river may deposit a sand bar along its banks. A river could carry sediment. It may be deposited on the sea floor at the river's mouth. This forms a delta. Waves may move sand from one spot to another.

CCC Cause and Effect

Wind can blow sand into piles. These make small dunes on a beach. Wind forms large sand dunes in places such as Death Valley. Glaciers leave piles of rocks where they melt. Sediments may be deposited inches or miles from where they were picked up. Some sediments are deposited in layers. These can turn into sedimentary rock over time.

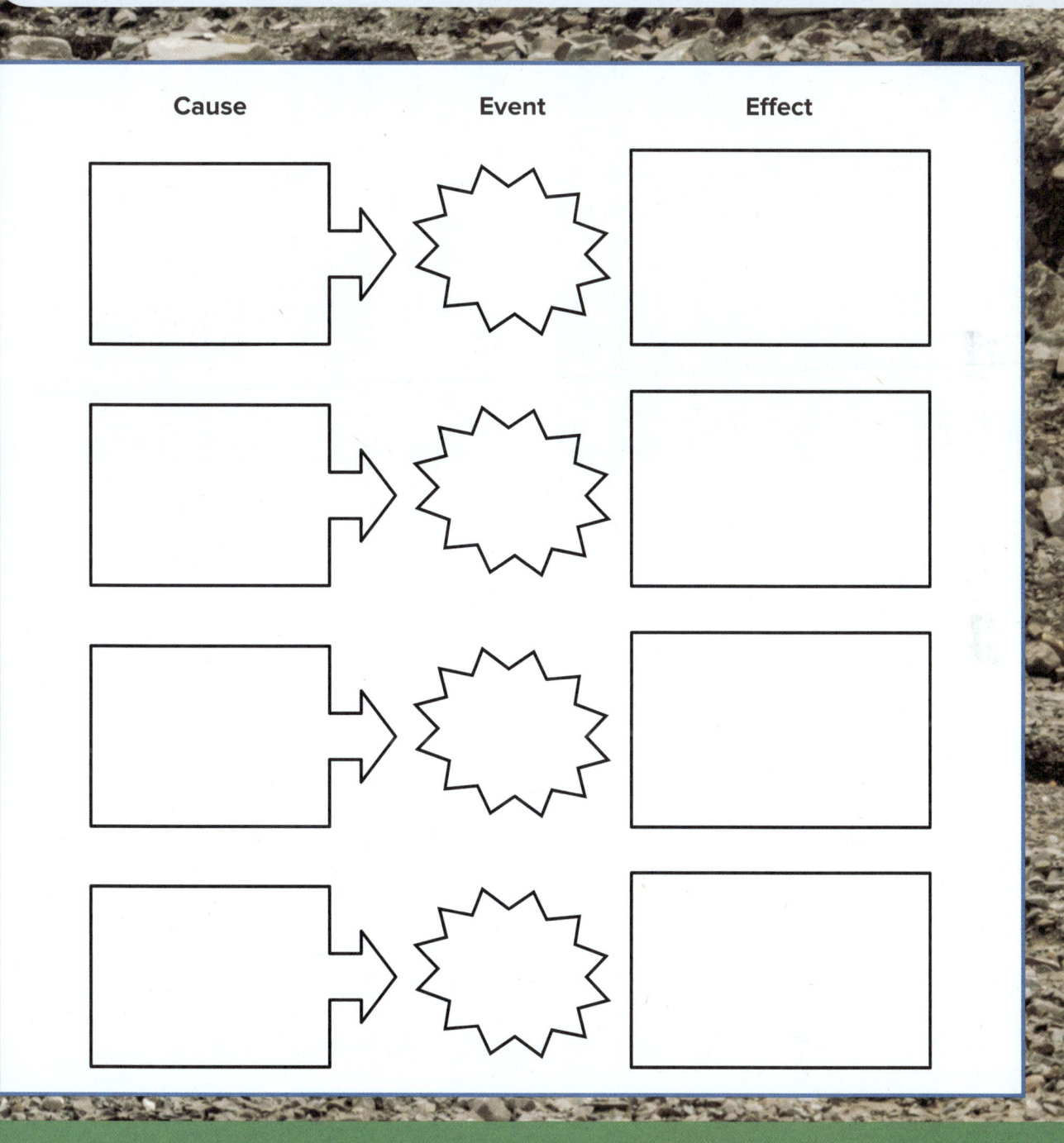

Cause	Event	Effect

Concept 3.1: Breaking Down and Moving Rocks | 35

3.1 | Learn — How do wind, water, and weather erode Earth's surface?

Activity 14
Evaluate Like a Scientist

Evidence of Change

Quick Code: us4522s

Look at the two images to help you **write** the definitions of *weathering*, *erosion*, and *deposition* in the table.

River Delta

Death Valley Dunes

Phenomenon	Definition
Weathering	
Erosion	
Deposition	

Concept 3.1: Breaking Down and Moving Rocks

3.1 | Share How do wind, water, and weather erode Earth's surface?

Activity 15
Record Evidence Like a Scientist

Quick Code: us4523s

Disappearing Sandcastles

Now that you have learned about wearing down and moving rocks, look again at the Disappearing Sandcastles image. You first saw this in Wonder.

Let's Investigate Disappearing Sandcastles

Talk Together

How can you describe disappearing sandcastles now?

How is your explanation different from before?

SEP Constructing Explanations and Designing Solutions

Look at the Can You Explain? question. You first read this question at the beginning of the lesson.

> ## Can You Explain?
>
> How do wind, water, and weather erode Earth's surface?

Now, you will use your new ideas about disappearing sandcastles to answer a question.

1. Choose a question. You can use the Can You Explain? question or one of your own. You can also use one of the questions that you wrote at the beginning of the lesson.

My Question

2. Then, use the graphic organizers on the next pages to help you answer the question.

Concept 3.1: Breaking Down and Moving Rocks

3.1 | Share — How do wind, water, and weather erode Earth's surface?

To plan your scientific explanation, first **write** your claim.

My claim:

_____
_____
_____
_____
_____

Next, **look** at your notes taken in Wonder, in the T-Chart for the activity Types of Weathering, in the table for the activity Forces That Shape Earth, and in the chart for the activity Badlands Map. Identify two pieces of evidence that support your claim:

Evidence 1

Evidence 2

Now, write your scientific explanation.

STEM in Action

Quick Code: us4524s

Activity 16
Analyze Like a Scientist

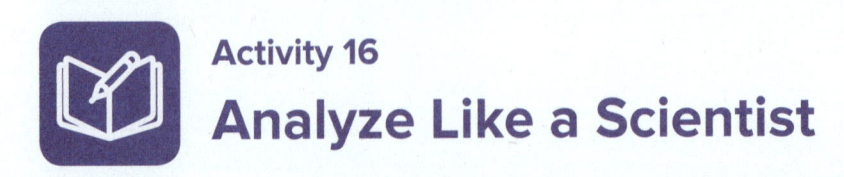

Careers and Erosion and Deposition

Read the text and **complete** the activities that follow.

Careers and Erosion and Deposition

Have you ever wondered why some rocks are smooth and some are jagged? Or why some rocks are large and others are small? If you have ever compared the shapes, sizes, and textures of rocks, you are playing the role of a geologist. A geologist is a scientist who studies rocks. Geologists study the history and structure of Earth and how it has changed over time.

In this video, geologist David Kring shares how he became interested in geology and the fascinating things he has discovered by studying the land around him.

By studying the rocks in the desert mountains, David has been able to uncover the history of the land and how weathering and erosion have caused massive changes in the landscape over millions of years.

Video

Mountains, Erosion, and Weathering

What have you learned in this video about the effects of erosion and weathering on the land as described by the geologist?

_____
_____
_____
_____
_____
_____
_____
_____

What are some other careers that study erosion and its effect on the land?

_____
_____
_____
_____
_____
_____
_____
_____

Working Like a Scientist

Explore the yard at school or home for rocks, or search the Discovery Education Techbook for images of rocks. **Select** three to four rocks that are different in shape and texture. **Write** down the characteristics of each rock (such as smooth, jagged, small, large) by using the following organizer.

Rock ID	Characteristics

What do you think caused these different characteristics?

3.1 | Share — How do wind, water, and weather erode Earth's surface?

Activity 17
Evaluate Like a Scientist

Review: Breaking Down and Moving Rocks

Think about what you have read and seen in this lesson. **Write** down some core ideas you have learned. **Review** your notes with a partner. Your teacher may also have you take a practice test.

Quick Code: us4525s

Talk Together

Think about what you saw in Get Started. Use your new ideas about breaking down and moving rocks to discuss the formation of the Grand Canyon.

Concept 3.1: Breaking Down and Moving Rocks

CONCEPT
3.2

Changing Landscapes

Student Objectives

By the end of this lesson:

- ☐ I can ask questions about the causes and stability of landforms that change slowly and quickly.
- ☐ I can provide evidence that weathering and erosion by wind, water, and ice cause changes on Earth's surface over time.
- ☐ I can develop a model that describes patterns in the formation of deltas and predicts where deltas are likely to form.
- ☐ I can describe the interactions between water and landforms in a watershed and between wind and sand dunes at the beach.
- ☐ I can use evidence from patterns in rock formations and data collected from stream table investigations to explain the changes in Earth's surface over time.

Key Vocabulary

- ☐ canyon
- ☐ delta
- ☐ dune
- ☐ glacier
- ☐ meander
- ☐ volcano

Quick Code: us4527s

Concept 3.2: Changing Landscapes

Activity 1
Can You Explain?

How are canyons formed?

Quick Code: us4528s

Concept 3.2: Changing Landscapes

3.2 | Wonder   How are canyons formed?

Activity 2
Ask Questions Like a Scientist

Quick Code: us4529s

Canyons

Observe the images. Then, **complete** the activity.

Let's Investigate Bryce Canyon

Let's Investigate Red Rock Canyon

Let's Investigate Slot Canyon

Let's Investigate Small Canyon

SEP Asking Questions and Defining Problems

What do you wonder about canyons? Think about how canyons look alike and different. **Write** three questions you have and **share** them with the class.

What

Why

When

Concept 3.2: Changing Landscapes | 53

3.2 | Wonder   How are canyons formed?

Activity 3
Investigate Like a Scientist

Quick Code: us4530s

Hands-On Investigation: School Landscape

In this investigation, you will take a class field trip outside. You will find and record evidence of change in a local landscape. You will use the evidence to create a map that shows different changes you found in the landscape.

Make a Prediction

Brainstorm what evidence of weathering, erosion, and deposition you could find in your schoolyard or a nearby park. What landforms would these processes create? **Write** your ideas in the Type of Landform boxes.

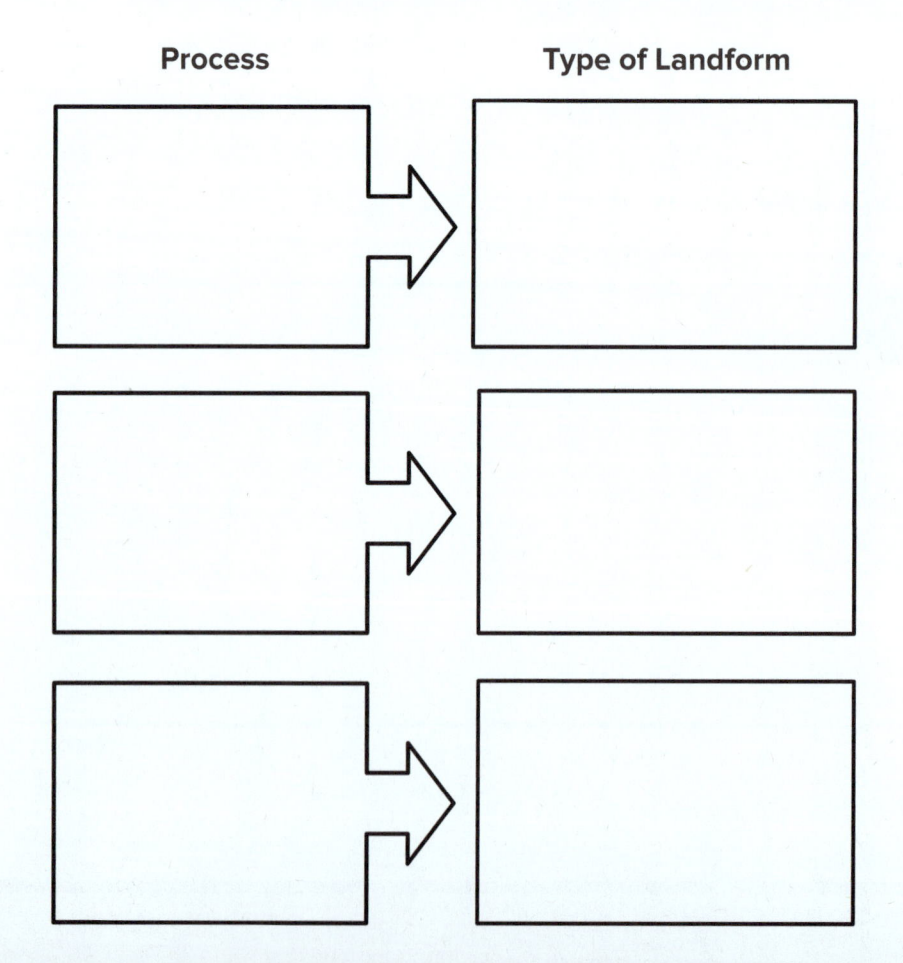

Process	Type of Landform

What materials do you need? (per group)

- Paper
- Pencils
- Clipboard
- Digital camera (optional)

What Will You Do?

1. Draw the main landscape areas in the provided space and label them.
2. Mark the areas where you observe a change and describe this change.
3. If you have a digital camera, use it to collect images of these areas.
4. When you have finished the map, combine it with the photos.

3.2 | Wonder How are canyons formed?

Think About the Activity

In this activity, you observed small models of landforms such as streams and hills. How would evidence of weathering, erosion, and deposition look different for larger landforms, such as canyons or mountains?

_____
_____
_____
_____
_____
_____

Explain why it might be useful to recognize signs of weathering, erosion, and deposition.

_____
_____
_____
_____
_____
_____

Compare your map to another group's map. Did you see different evidence? Is there anything they have that you would put on your map if you did it again?

_____

_____

_____

_____

_____

Concept 3.2: Changing Landscapes | 57

3.2 | Wonder How are canyons formed?

Activity 4
Observe Like a Scientist

Shaping the Landscape

Quick Code: us4531s

Look at the images of different landscapes at different scales. **Look** for similar and different features in the images. Then, **answer** the questions that follow.

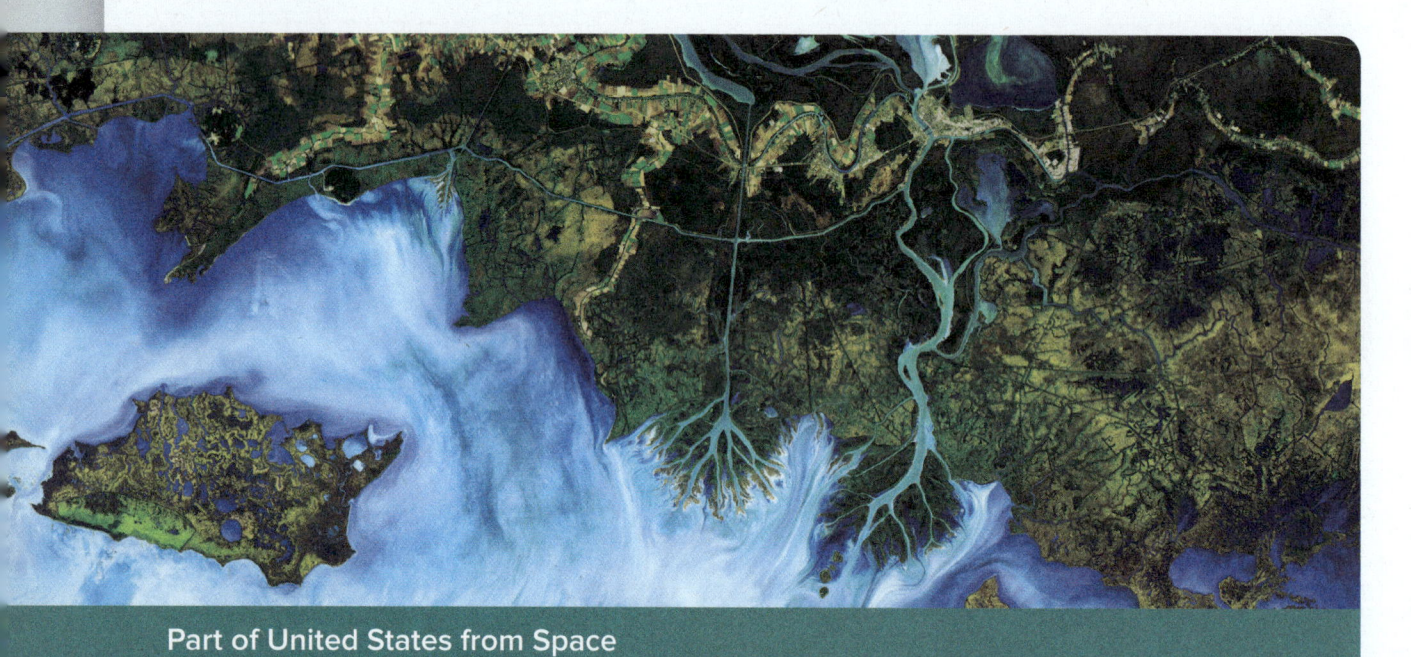

Part of United States from Space

Grand Canyon

Greenland

How do your observations from the School Landscape activity compare to the images?

_____

_____

What are some advantages of looking at Earth's surface at different scales?

_____

_____

_____

Concept 3.2: Changing Landscapes | 59

3.2 | Wonder — How are canyons formed?

Activity 5
Evaluate Like a Scientist

Quick Code: us4532s

What Do You Already Know About Changing Landscapes?

How Did It Form?

Look at the gully in the image. Then, **write** your answer to the following questions.

Gully

How do you think the gully was formed?

_____

_____

_____

_____

How can this observation help predict future changes?

_____
_____
_____
_____

Landforms

Look at the images of landforms. **Write** each of the following labels below the landform it describes.

| Canyon | Dunes | Mountain | Plains | Plateau | Valley |

_____   _____   _____

_____   _____   _____

Concept 3.2: Changing Landscapes | 61

3.2 | Wonder — How are canyons formed?

Explain

Write an explanation of the formation process for each landform in the table.

Landform	Formation Process
Canyon	
Dunes	
Mountain	
Plains	
Plateau	
Valley	

Concept 3.2: Changing Landscapes | **63**

3.2 | Learn How are canyons formed?

How Do Landscapes Change?

Activity 6
Observe Like a Scientist

Visual Walkabout

Quick Code: us4533s

Look at the gallery of images. **Look** for characteristic features of the landforms in the images. **Write** questions about the landforms in the organizer provided. Then, **write** possible answers.

Image	Your Questions	Your Answers

64

Image	Your Questions	Your Answers

CCC Cause and Effect

Concept 3.2: Changing Landscapes

What Sorts of Landforms Are Shaped by Water and Ice?

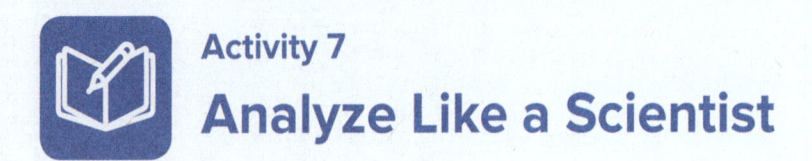

Activity 7
Analyze Like a Scientist

Canyon Formation

Quick Code: us4534s

Read the following statements and **check** the boxes to indicate if you agree or disagree with each statement.

Agree	Disagree	Statement
		The bigger the stream, the more erosion it causes.
		Rivers erode rocks and can form valleys and canyons.
		Canyon walls are not very tall and have gentle slopes.
		A canyon is a type of valley.
		Rivers can change a landform very slowly.
		Fast-moving rivers can cause a lot of erosion.

Now, **read** the text. After you read, **review** your answers and change them, if necessary.

CCC Patterns

Canyon Formation

Gravity pulls rainwater downhill where it gathers into small streams. These streams join one another and form bigger streams. Big streams or rivers cause more erosion than little streams. Rivers form valleys as they erode deeper into the landscape. Many valleys are formed by rivers. The shape of the valley formed depends upon several factors, including the types of rocks present and the speed, age, and size of the river.

As rivers drain, they create a host of different landforms on the landscape. Have you ever heard of the Grand Canyon? It is very large and steep. If you look down into it, you can see many layers of rock. In several places the canyon walls are almost vertical. Looking down at the very bottom, you can see the Colorado River. Canyons are special types of valleys with steep sides. You can probably guess how this canyon formed. Over a long period of time, the river eroded the rock and cut farther and farther down into the rock. Because the river was going down a steep grade, it was moving fast and had a lot of energy. It could erode a lot of **sediment** and carry it away. This process took many millions of years.

3.2 | Learn   How are canyons formed?

Activity 8
Observe Like a Scientist

Canyons and Valleys

Watch the videos and **look** for patterns in how valleys and canyons are formed. Then, **answer** the questions.

Quick Code: us4535s

Video — The Grand Canyon

Video — Valleys and Canyons

CCC Patterns

How was the Grand Canyon formed?
_____
_____
_____

Why do you think the land around the Grand Canyon was not eroded at the same rate?
_____
_____
_____

What features are characteristic of a canyon?
_____
_____
_____

Concept 3.2: Changing Landscapes

3.2 | Learn How are canyons formed?

Activity 9

Evaluate Like a Scientist

Delta Formation

Quick Code: us4536s

Read the text. Then, **complete** the activity that follows.

Unlike valleys and canyons, deltas are not formed by erosion. Deltas are formed by a process called deposition. They form when streams or rivers carrying lots of silt slow down. When the river slows, most of the silt drops out of the water. Deltas are flat, fan-shaped deposits of sediment. The largest **delta** in the United States is the Mississippi River delta. It formed where the Mississippi River enters the Gulf of Mexico. Most deltas form where flowing water enters still water. This could be a large river entering the sea or a mountain stream entering a lake. The key is that deltas form where water loses energy and drops the sediment it's carrying. Large wetlands form in deltas. The wetland plants are partly responsible for slowing down the water and their roots help trap sediments. This increases the rate of deposition.

SEP Developing and Using Models

Now, **look** at the map that shows a river flowing through a lake and then into an ocean. Work with a partner to **draw** crosses on the map where you think deltas will form.

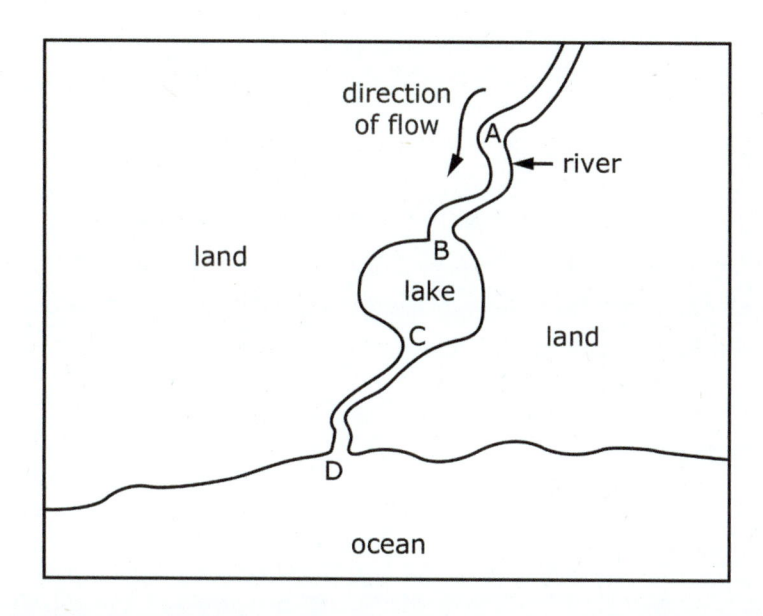

Explain why you marked these locations.

_____
_____
_____
_____
_____
_____
_____
_____

Concept 3.2: Changing Landscapes | 71

3.2 | Learn How are canyons formed?

Activity 10

Investigate Like a Scientist

Hands-On Investigation: Using a Stream Table to Model Valley Landforms

Quick Code: us4537s

In this investigation, you will use a stream table to model and explore how streams are formed and how the steepness of the land can affect the shape of the stream. You will change the steepness of the stream table to learn how this variable affects the shape of the stream.

Make a Prediction

First, **write** your predictions in the chart.

Question	My Prediction
What factors determine how deep and wide a stream is?	

CCC Patterns

What materials do you need? (per group)

- Large bin, with lid, 2
- Duct tape
- Sand, fine
- Stream table—short ruler or piece of wood
- Empty 2-liter bottle
- Markers
- Foam cup, 6 oz
- Toothpick
- Stopwatch
- Metric ruler
- Protractor
- Blocks or books for propping

What Will You Do?

First, set up the stream table.

1. Drill a hole in the corner of a plastic box and cover it with duct tape.
2. Fill the box with sand.
3. Leave the part with the covered hole empty.
4. Smooth the surface of the sand.
5. Draw a horizontal line near the top of a bottle.
6. Use a toothpick to poke a small hole in the bottom of a foam cup.
7. Place the bucket under the hole of the box and remove the duct tape.

Next, do the experiment.

1. Add water to the empty bottle up to the fill line.
2. Hold the cup over the end opposite the drilled hole.
3. Have your partner pour from the bottle into the cup for 2 minutes or until the bottle is empty.

Concept 3.2: Changing Landscapes

3.2 | Learn   How are canyons formed?

4. Measure and record how deep and large the stream is.
5. Write a brief description of the stream in the table below.
6. Tilt the stream table to drain the water.
7. Arrange the sand as it was at the beginning.
8. Repeat the experiment.

Experiment	Stream Description

Think About the Activity

What patterns did you observe in how steepness affected stream depth and width?

_____
_____
_____
_____

How do you explain the patterns you observed?

_____
_____
_____
_____
_____

How does your model compare to streams and rivers in nature?

_____
_____
_____
_____
_____

What results would you predict for testing other variables with your stream table?

_____
_____
_____
_____

Concept 3.2: Changing Landscapes | 75

Activity 11
Analyze Like a Scientist

Quick Code: us4538s

Erosion by Glaciers

Read the text. As you read, **record** evidence to support the claim "Landscapes change over time" in the organizer. Also include evidence from the previous activities. Use the organizer to explain how your evidence can support this claim.

Erosion by Glaciers

Valleys can be eroded by ice as well as water. **Glaciers** are made from ice that forms when snow does not melt. They are found in polar regions and on high mountains. As glaciers build up, they become very heavy. Glaciers move due to Earth's gravity—they flow downhill.

The movement of glaciers can cause erosion that changes Earth's surface. Many valleys, mountains, and lakes were formed or altered by glaciers. For example, about 18,000 years ago during the last Ice Age, regions of North America as far south as what is now the state of Indiana were covered in glaciers. As the glaciers moved, they carved many features into the landscape, including rivers, lakes, and hills. Landforms made by glaciers have special characteristics that can be identified by scientists. Scientists can use these landforms to figure out where glaciers existed in the past.

Yosemite Valley

The claim: Landscapes change over time.

Evidence I found: Record all the evidence you gathered from the text and the previous activities.

The claim is true because:

| SEP | Constructing Explanations and Designing Solutions |

Concept 3.2: Changing Landscapes | 77

How Does Wind Create Landforms?

Activity 12
Analyze Like a Scientist

Quick Code: us4539s

Wind Erosion

Read the text and **look** at the images. **Draw** the outline of one image that represents the main idea of the text. Then, **fill in** the outline with facts you learned from the text and images.

Wind Erosion Creates Landforms

Wind Erosion

When wind blows across the land, it picks up sand and other rock particles and carries it along. When this flying sediment hits a rock, it wears down that rock like a sandblaster. This carves the rock into strange shapes.

Some landforms are created by erosion and deposition processes at the same time. Have you ever been to a beach or to a sandy desert? What landforms could these two very different environments have in common? Sand dunes, of course!

CCC Cause and Effect

As the name implies, these landforms are made of wind-blown sand. You usually see dunes in groups, and they may cover a large area. They can be hundreds of meters tall.

Dunes are interesting because they are constantly moving. When wind blows across a dune, sand grains erode away from the side the wind is coming from. The grains bounce along, up the slope of the **dune**. When they reach the top, they enter still air behind the dune and roll down the other side. Dunes form because the wind is not strong enough to carry away the grains.

Sand Dunes: Shifting Landscapes

Concept 3.2: Changing Landscapes | 79

3.2 | Learn — How are canyons formed?

Activity 13
Investigate Like a Scientist

Quick Code: us4540s

Hands-On Investigation: Sand Shifters

In this investigation, you will create a model to explore how wind creates sand dunes and think about the factors that cause sand dunes to form.

Make a Prediction

First, **write** your predictions in the chart.

Question	My Prediction
How do sand dunes form?	
Why do sand dunes form in some locations and not others?	

SEP Developing and Using Models

What materials do you need? (per group)

- Aluminum foil pan, 13 × 9 × 2
- Broom and dustpan
- Copy paper box lids (for catching extra blowing sand if activity is done inside), 3
- Spray bottle
- Cooking oil spray (can be shared as a class)
- Plastic straws
- Colored pencils
- Safety goggles (per student)
- Sand

What Will You Do?

1. Fill three pans with sand. Place a small rock in each container.
2. Think how to produce a build-up of sand in one place with the provided materials.
3. Write your predictions in your science journal.
4. Explore what happens when you use the straw to blow sand.
5. Record your observations in your science journal.

Concept 3.2: Changing Landscapes

3.2 | Learn — How are canyons formed?

Think About the Activity

How does the wind affect the sand?

_____
_____
_____
_____

What patterns did you see in the sand?

_____
_____
_____
_____

Compare your findings with other groups. Explain how they were similar or different.

_____
_____
_____
_____
_____

How Does Deposition Make Layers of Rock?

Activity 14
Investigate Like a Scientist

Quick Code: us4541s

Hands-On Investigation: Modeling Deposition and Rock Formation

In this investigation, you will use a stream table and different types of sediment to model how erosion and deposition can form rock layers. You will also use the stream table to understand how fossils can become a part of rock layers.

Make a Prediction

First, **write** your predictions in the chart.

Question	My Prediction
Why can we see layers of different rock in landforms like canyons?	
What can we learn from these layers of rock?	

SEP Developing and Using Models
CCC Cause and Effect
CCC Patterns

Concept 3.2: Changing Landscapes | 83

3.2 | Learn How are canyons formed?

What materials do you need? (per group)

- Foam cup, 6 oz
- Sand
- Soil, potting
- Peat
- Leaves
- Large bin, with lid, 2
- Empty 2-liter bottle
- Duct tape

What Will You Do?

Deposition

1. Remove the rubber stopper from the stream table and place a container under the hole.
2. Pour a layer of sand into the stream table.
3. Fill the bottle from the stream table up to the fill line and pour it into the foam cup.
4. Let the water drain out, but do not shake the table.
5. Repeat steps 2–4 with a layer of soil on top of the sand already present.
6. Repeat steps 2–4 with a layer of peat on top of the soil present.

Fossilization

1. Place leaves on top of the layers that formed at the bottom of the hill in the Deposition part.
2. Pour a layer of sand on top of the hill.
3. Fill the bottle from the stream table up to the fill line and pour it into the foam cup.
4. Record your observations and draw sketches of the experiment in this table.

Action	Observations	Sketch
Pour water over three layers of sediment.		
Pour water over the sand with the leaves at the bottom.		

Concept 3.2: Changing Landscapes

3.2 | Learn — How are canyons formed?

Think About the Activity

How do rock layers provide evidence for how an area changes over time?

_____
_____
_____
_____

How does your model help to explain why rock layers often differ in color?

_____
_____
_____

Soil along the banks of a fast-moving river (like one coming down a mountain) does not have as many different nutrients as soil around a slow-moving river (like one in the middle of a flat valley). What evidence from your investigation could explain this pattern?

_____
_____
_____
_____

What causes the differences in how canyons look? Explain your thinking using evidence from the activity.

_____
_____
_____
_____

Activity 15
Analyze Like a Scientist

Rock Layers of Zion National Park

Quick Code: us4542s

Read the text. When you finish reading, **underline** the main idea.

Rock Layers of Zion National Park, Part 1

Zion National Park in southern Utah is known for its dramatic cliffs, canyons, and rivers. The rugged terrain is about 3,000 feet above sea level. Visitors might think that Zion has always looked as it does now. However, the layers of rock tell a different story.

When rocks weather and erode, they form sediments. These sediments become deposited in layers. The layers may include fossils of plant and animal life that existed at the time. Scientists look at the rock layers to determine what the area looked like long ago.

CCC Patterns

Concept 3.2: Changing Landscapes | 87

View the image.

Rock Layers of Zion National Park

How does this image relate to the main idea you underlined?

_____

_____

_____

_____

Now, **read** the rest of the text. As you read the descriptions of each of Zion's rock layers, **think** about how we know the environment changed over time. Then, **answer** the questions that follow.

Rock Layers of Zion National Park, Part 2

Kaibab Formation

Around 270 million years ago, all the land on Earth was part of the supercontinent, Pangaea. The area of the modern park was at sea level near the equator. Much of the layer is limestone and siltstone. These minerals give clues about what the area looked like. They show that the area was a shallow tropical sea and coastal flats. You can find marine fossils here. These fossils include types of clams, coral, and trilobites.

Moenkopi Formation

By 240 million to 250 million years ago, the area that became Zion National Park had changed. It was covered in slow-moving rivers and broad tidal flats. The mudstone and sandstone we find in these layers have ripple marks in them. These patterns show that water and waves moved the sediments. The presence of limestone and gypsum is evidence of an evaporating tidal flat. Fossil footprints of early reptiles and amphibians show which animals inhabited the area.

Chinle Formation

By 210 million to 225 million years ago, a large river system covered the area. The rivers deposited many types of sediment. Fast rivers laid down big pieces of rock, such as gravel. Slower rivers laid down mud and ash from nearby **volcanoes**. This layer also contains massive petrified conifer tree trunks and fossils of early dinosaurs.

Rock Layers of Zion National Park, Part 2 cont'd

Moenave Formation/Springdale Sandstone/Kayenta Formation

The Moenave Formation was laid down 210 million to 195 million years ago. One part of the Moenave Formation has clear fossils of dinosaur footprints around an ancient lake. Fossils of fish and ancient plants also occur in this layer.

The Springdale Sandstone layer is a part of the Kayenta Formation. It was deposited by a river that connected to the river that dropped the Kayenta and Moenave formations. Today, the Moenave Formation is eroding, causing pieces of the Springdale Sandstone to fall off.

The Kayenta Formation was laid down 195 million to 185 million years ago. It is mostly mudstone, made from deposits in river floodplains and small lakes. It also has sandstone. Sandstone results from deposits left behind by the streams. Dinosaur tracks occur in this layer.

Navajo Sandstone

By 185 million to 180 million years ago, this area was a desert. It was part of the largest known sand desert in Earth's history! Scientists know this because these rock layers are sandstone that shows cross-bedding. Cross-bedding happens when wind blows across sand dunes. The layers of sediment are therefore laid down at an angle.

Temple Cap Formation

This layer is made of sandstone, mudstone, and limestone. The variety of rock types indicates that a shallow sea was developing in the area as the layer was forming 175 million to 170 million years ago.

Carmel Formation

The Carmel Formation is mostly made of limestone formed 170 million to 165 million years ago. This mineral contains marine fossils, including the shells of clams and snails. Similar animals are found in oceans today. The layer also includes mudstone, sandstone, and gypsum. This evidence shows that this area was a warm, shallow sea with sandy deserts around it.

Volcanic Rocks

Between 10 million to 100,000 years ago, this area experienced dramatic change. Nearby volcanoes erupted. Their lava spread over the surface. You see this as a thin layer of volcanic rock.

Now, use the information in the passage to **answer** the questions.

How do scientists use patterns in rock formations, and fossils in rock formations, to explain the changes in a landscape over time?

Consider the Kaibab and Moenkopi formations. The Narrows is a popular hiking spot in Zion National Park. This section of the Virgin River is very narrow. It is sometimes only 20 feet wide. The rock walls on either side of the river are around one thousand feet tall. Hikers wading in the river through the gorge may notice that the rock walls on either side of them have matching layers. What evidence in these layers explains how the area changed from a shallow sea to a river system? According to the evidence, how did the landscape of the Narrows change over time?

_____
_____
_____
_____
_____
_____
_____
_____
_____
_____
_____
_____
_____

Concept 3.2: Changing Landscapes

3.2 | Learn How are canyons formed?

Activity 16
Evaluate Like a Scientist

Quick Code: us4543s

Describing Landforms, How Landforms Are Formed, and How Quick Is Erosion?

Describing Landforms

Write the following words in the blanks to correctly complete the passage.

| Lower | Deltas | Sand dunes |
| Higher | Canyons | |

Canyons, deltas, and sand dunes are some of Earth's landforms. _____ are deep valleys with steep sides. _____ are fan-shaped landforms where rivers enter lakes or oceans. _____ are hills that are made of sand. Rivers always flow from _____ elevations to _____ elevations.

CCC Patterns

CCC Cause and Effect

How Landforms Are Formed

In the columns, **write** how each landform is caused. There can be more than one cause for each landform.

Erosion   Water   Wind   Ice

	Canyon	Delta	Sand Dune
Causes			

How Quick Is Erosion?

Write "quickly" or "slowly" in the blanks to correctly complete each sentence.

During a storm or a rockslide, erosion can happen _____.

In general, erosion happens _____.

Concept 3.2: Changing Landscapes

3.2 | Share How are canyons formed?

Activity 17
Record Evidence Like a Scientist

Quick Code: us4544s

Canyons

Now that you have learned about changing landscapes, look again at the images of canyons. You first saw these in Wonder.

Let's Investigate Bryce Canyon

Let's Investigate Red Rock Canyon

Let's Investigate Slot Canyon

Let's Investigate Small Canyon

Talk Together

How can you describe canyons now?
How is your explanation different from before?

SEP Constructing Explanations and Designing Solutions

Look at the Can You Explain? question. You first read this question at the beginning of the lesson.

> ### Can You Explain?
>
> How are canyons formed?

Now, you will use your new ideas about the formation of canyons to answer a question.

1. Choose a question. You can use the Can You Explain? question or one of your own. You can also use one of the questions that you wrote at the beginning of the lesson.

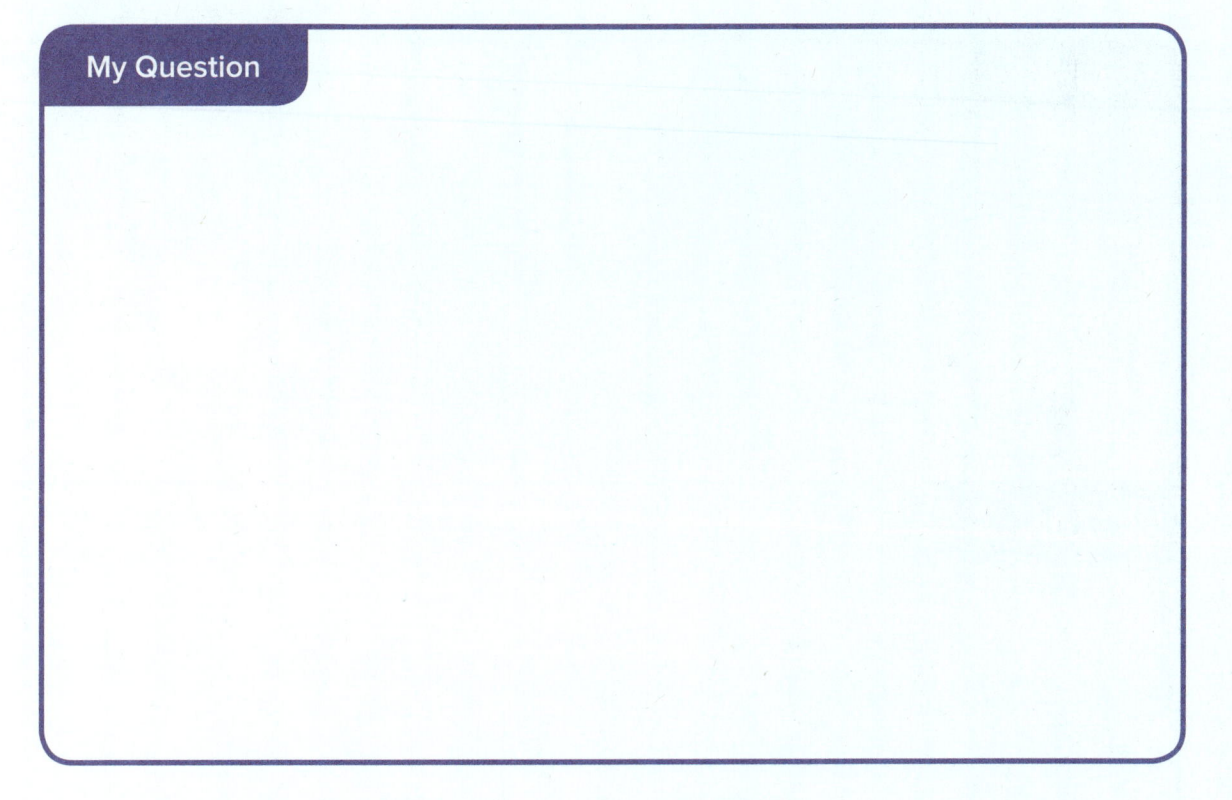

My Question

2. Then, use the graphic organizers on the next pages to help you answer the question.

Concept 3.2: Changing Landscapes

3.2 | Share — How are canyons formed?

To plan your scientific explanation, first **write** your claim.

My claim:

_____
_____
_____
_____

Next, **look** at your notes taken in the tables for the activities Explain and Visual Walkabout and for the answers to the activities Canyon Formation and Canyons and Valleys. **Identify** two pieces of evidence that support your claim:

Evidence 1

Evidence 2

Now, **write** your scientific explanation.

STEM in Action

Quick Code: us4545s

Activity 18
Analyze Like a Scientist

It's a Snap: Photographers, Photos, and Landforms

Read the text and **watch** the video. Then, **complete** the activities that follow.

It's a Snap: Photographers, Photos, and Landforms

Photographers capture stories with their photos. What does the photo below tell you? Well, you know just by looking at it that a photographer took this photo on a sunny day near a shore made of a type of black rock. (Perhaps it is lava!) The photo can also tell you of past events. For example, what might have caused the archway to form? Maybe you will conclude that the constant ocean waves pounding on the rock caused it to change its shape.

Arch along Lava Shore Caused by Wave Erosion

100 | Discovery Education

If the photographer returned to this location twenty years from now and took a new photo, what might the photo show? What story would it tell you? Photos help us to understand events, such as how Earth's surface changes.

Cameras are the main tool that photographers use. In the past, cameras were used only to take still pictures. Today, technology allows cameras to do many amazing things. For example, photographers can use a technique called time-lapse photography to take many pictures of the same place for a certain period of time. It can be a day, a month, or even a year! The still photos are placed in order to show the changes that happen in that one place.

Geographers and other scientists observe the time lapse sequences to draw conclusions about how the landforms in that area were made. They can make predictions of what they would look like in the future.

How can you use photos and photography to learn how landforms are formed?

Watch the video to see an example of time-lapse photography. Then, **answer** the questions that follow.

Video

Wind Erosion

Concept 3.2: Changing Landscapes | 101

Using Photography

Using what you have learned, **answer** the questions.

Based on what you have just watched, what caused the sand in the desert to change its location?

_____

_____

_____

Would you be able to watch a landform like a canyon or a delta form using time-lapse photography?

_____

_____

_____

Designing Technology

Because the shape of the desert shifts and changes, people can easily get lost. **Imagine** you are in charge of a group of scientists building technology to help find people who are lost in the desert. What tool would you build, and why? **Describe** the tool and its purpose in the space below.

3.2 | Share — How are canyons formed?

Activity 19
Evaluate Like a Scientist

Quick Code: us4546s

Review: Changing Landscapes

Think about what you have read and seen. What did you learn?

Write down some core ideas that you have learned. **Review** your notes with a partner. Your teacher may also have you take a practice test.

_____
_____
_____
_____
_____
_____
_____
_____
_____
_____
_____
_____
_____

SEP Obtaining, Evaluating, and Communicating Information

Talk Together

Think about what you saw in Get Started. Use your new ideas about changing landscapes to discuss the formation of the Grand Canyon.

Concept 3.2: Changing Landscapes | 105

CONCEPT
3.3

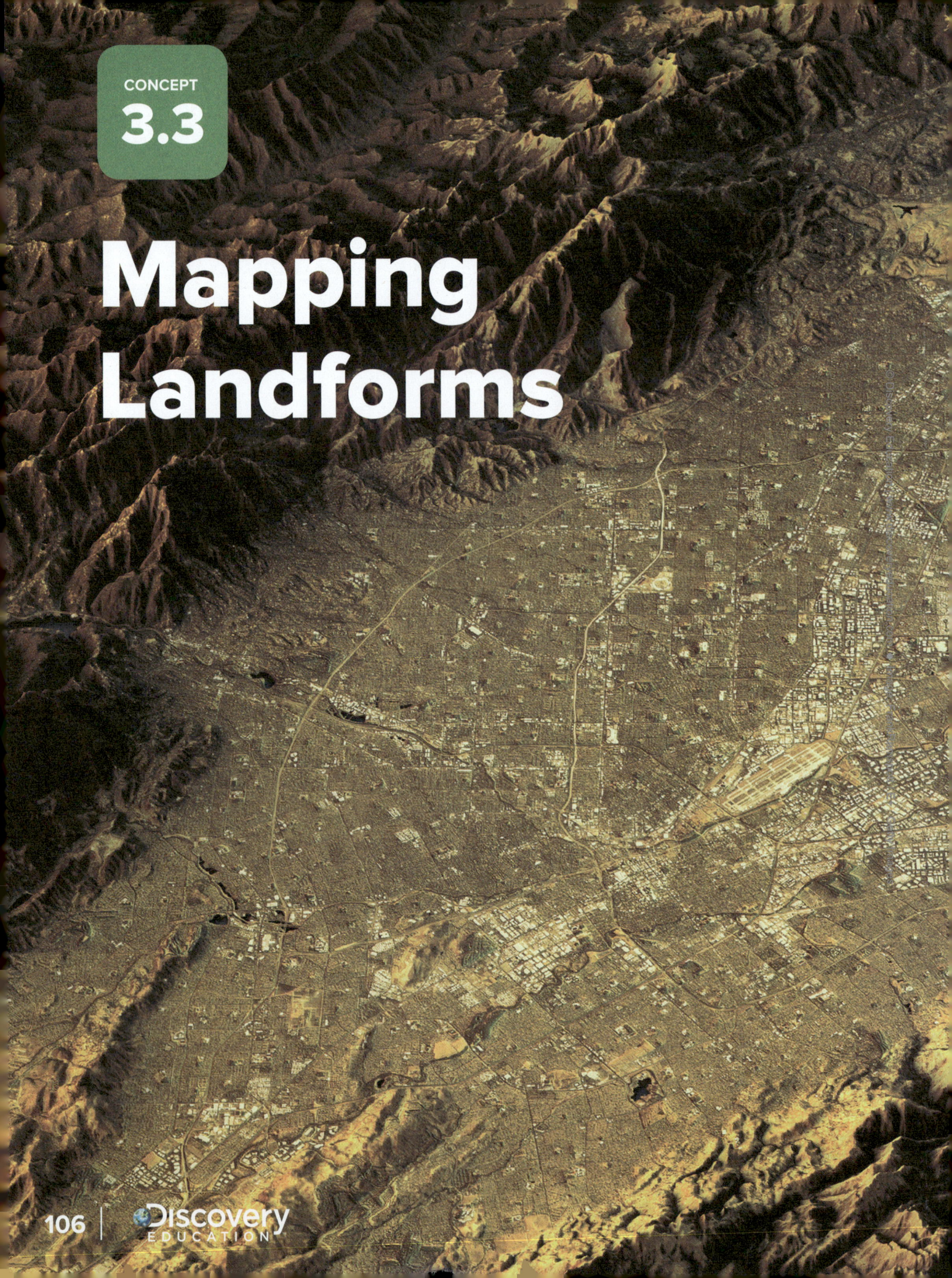

Mapping Landforms

Student Objectives

By the end of this lesson:

☐ I can analyze and interpret data from maps to describe patterns on Earth's surface on small and large scales.

Key Vocabulary

☐ landform
☐ map
☐ mountain
☐ satellite
☐ topographic map

Quick Code: us4548s

Concept 3.3: Mapping Landforms

Activity 1
Can You Explain?

How can maps be used to discover patterns in Earth's landscape features?

Quick Code: us4549s

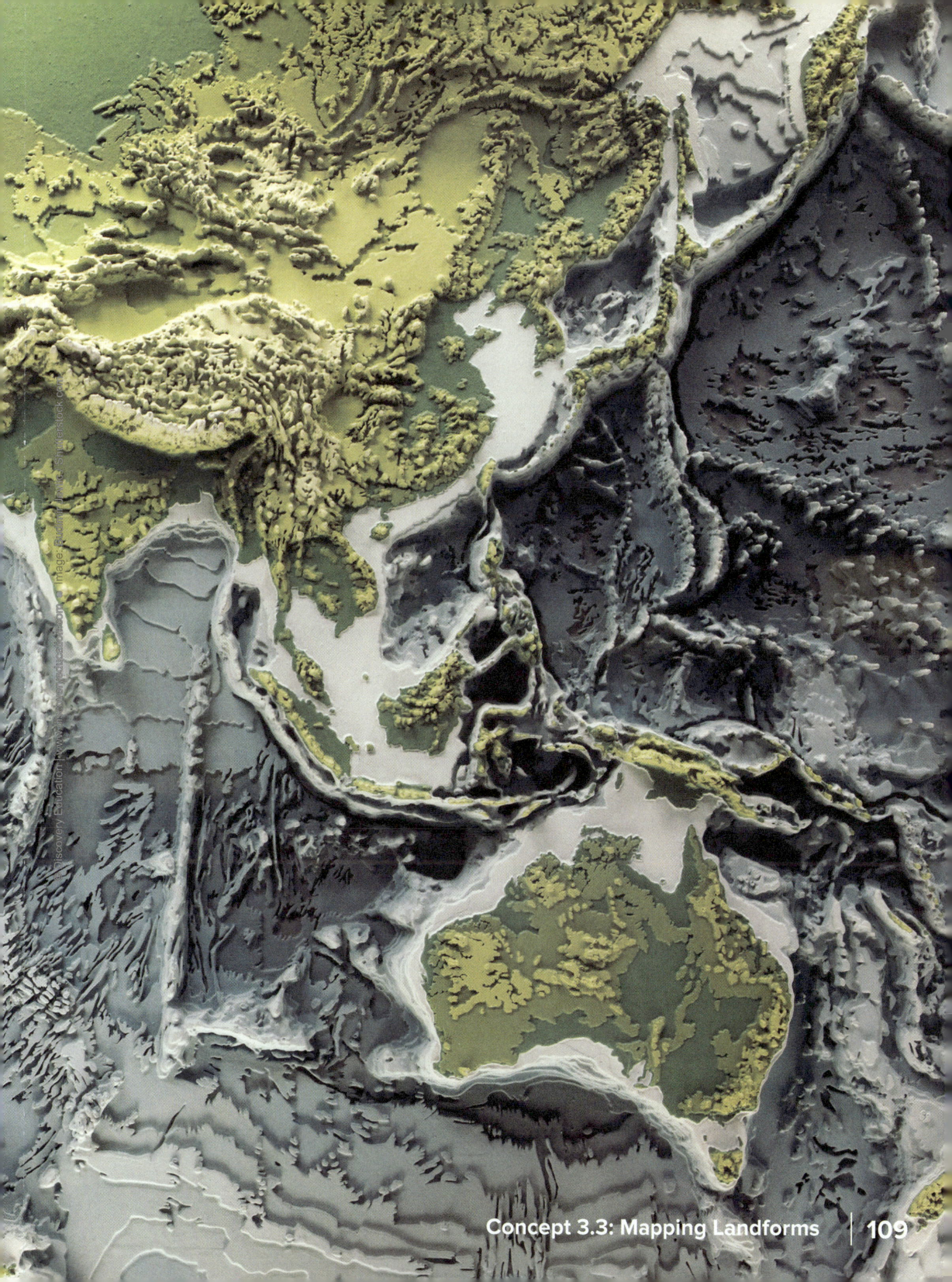

Concept 3.3: Mapping Landforms | 109

3.3 | Wonder
How can maps be used to discover patterns in Earth's landscape features?

Activity 2
Ask Questions Like a Scientist

Satellite Image of an Ocean

Look at the satellite image map. **Mark** areas on the map that represent high and low elevation. Then, **complete** the activity.

Quick Code: us4550s

Let's Investigate a Satellite Image of an Ocean

SEP Asking Questions and Defining Problems

What do you wonder about the landforms we can observe from satellite images? Think about the maps and images used to explore landforms under the oceans and on land. **Write** three questions you have and **share** them with the class.

I wonder...

I wonder...

I wonder...

Concept 3.3: Mapping Landforms

Activity 3
Analyze Like a Scientist

Quick Code: us4551s

Under the Ocean

Read the text and **view** the map. Then, **answer** the questions that follow.

Under the Ocean

Oceans stretch thousands of miles among the seven continents. The ocean's average depth is about 3,400 meters, or more than 2 miles. In fact, the continent that you are standing on is not just an island. You are on the peak of a **mountain** more than 2 miles high!

The features of the shoreline are visible as you stand on the beach, but what do you think the ocean floor looks like? Just as there are breathtaking continental **landforms**, there are also landforms on the ocean floor.

Map of the Ocean Floor

CCC Scale, Proportion, and Quantity

What landforms would you expect to see on the ocean floor?

I expect of sand costally

How are landforms on the ocean floor similar to and different from continental landforms?

Why is it important to know what is on the ocean floor?

Concept 3.3: Mapping Landforms

3.3 | Wonder
How can maps be used to discover patterns in Earth's landscape features?

Activity 4
Observe Like a Scientist

The Ocean Floor

As you **watch** the video, **look** for landforms on the ocean floor.

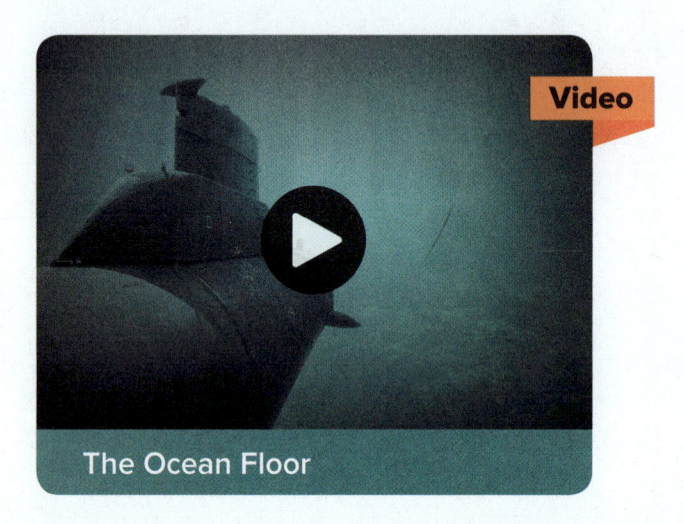

Quick Code: us4552s

[Video: The Ocean Floor]

Talk Together

Now, talk together about how scientists have mapped the ocean floor.

That use superery ecleorcanshen or soner, on water roset and sadeflint

Activity 5

Evaluate Like a Scientist

Quick Code: us4553s

What Do You Already Know About Mapping Landforms?

Where's Walter?

Walter is checking out different continental landforms. **Read** each description and determine where Walter is. **Write** the correct landform from the list next to its description. Not all landforms will be used.

| Plains | Mountains | Plateaus | Valleys | Hills |

Description	Which Landform Is It?
Walter is standing on a low area of land looking at the higher level of land to his left and right.	
Walter feels a sense of accomplishment as he is now standing on a peak.	
Walter is enjoying riding his bike on a flat area with some gentle slopes.	
Walter is looking at the land below as he walks on a flat surface at a high elevation.	

CCC Patterns

Concept 3.3: Mapping Landforms | 115

3.3 | Wonder — How can maps be used to discover patterns in Earth's landscape features?

Landform Elevation

Examine the shaded relief map of Egypt. **Look** at the landforms labeled A through D. Then, **rank** the landforms from highest to lowest elevation, with A being the highest and D being the lowest.

Egypt Map

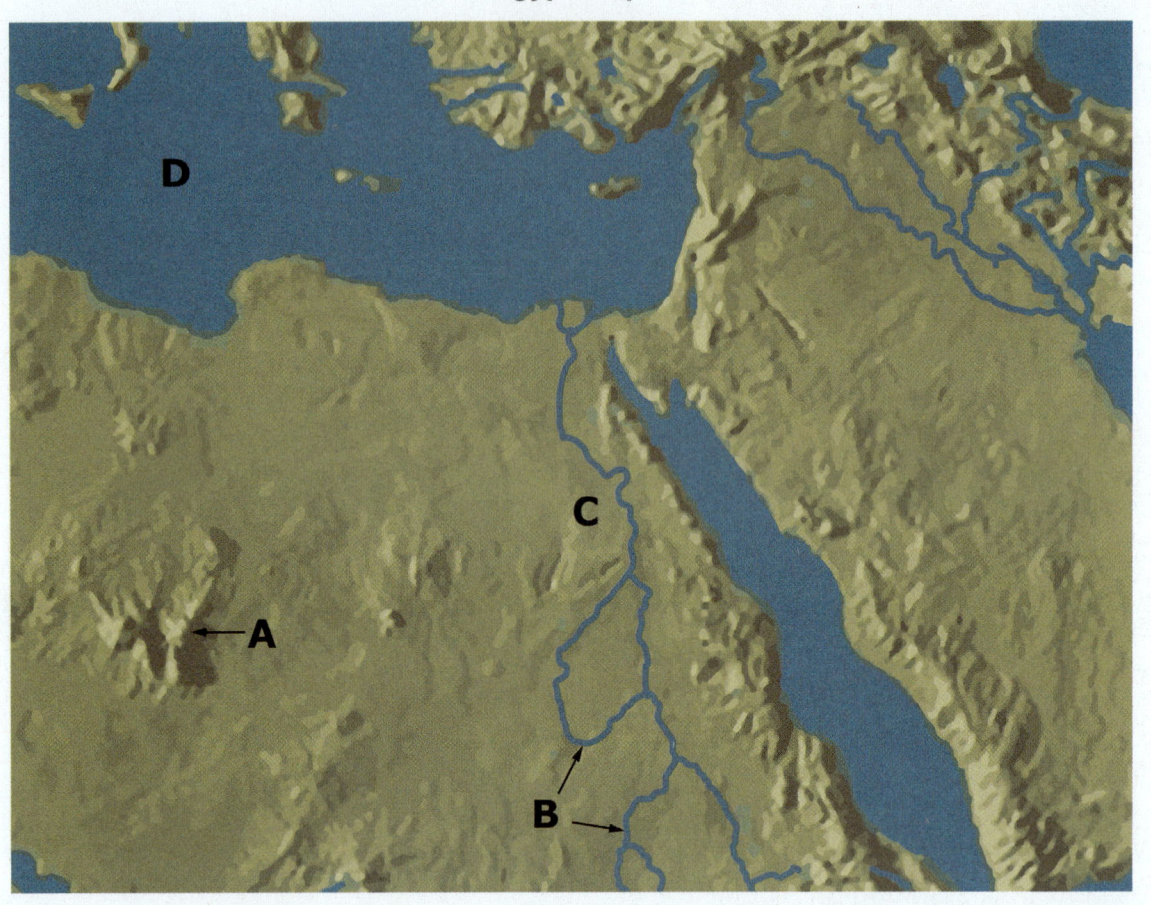

	Landform (Letter)
Highest Elevation	
Second-Highest Elevation	
Second-Lowest Elevation	
Lowest Elevation	

Concept 3.3: Mapping Landforms | 117

How Can Maps Be Used to Provide Information about Landforms?

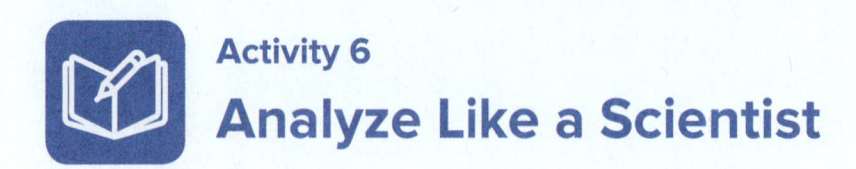

Activity 6
Analyze Like a Scientist

Quick Code: us4554s

Information on a Map

Read the text. As you read, **underline** the information that maps provide.

Information on a Map

When was the last time you used a **map**? What did you use it for? Maps not only help you get where you need to go, but also give important information about landforms. For example, maps can show the shape, size, and location of landforms. Additionally, maps can provide important information about Earth's landforms, such as how high or how low the landform is, or what natural resources are found in the landforms. Maps that show the landscape are called physical or **topographic maps**. Many physical maps are colored. Color is used to show water and the height of different landforms like mountains and hills.

Symbols are used to create maps. One such symbol is the contour line. Contour lines are lines whose closeness shows the slope, or how steep a landform is. The closer the contour lines, the steeper the hill.

Physical Map of the United States and Canada

Concept 3.3: Mapping Landforms | 119

3.3 | Learn — How can maps be used to discover patterns in Earth's landscape features?

Activity 7
Observe Like a Scientist

Comparing Map Features

As you **view** the video and images, **look** for the information they provide about Earth's surface. Then, use the Venn Diagram to **compare** the three maps.

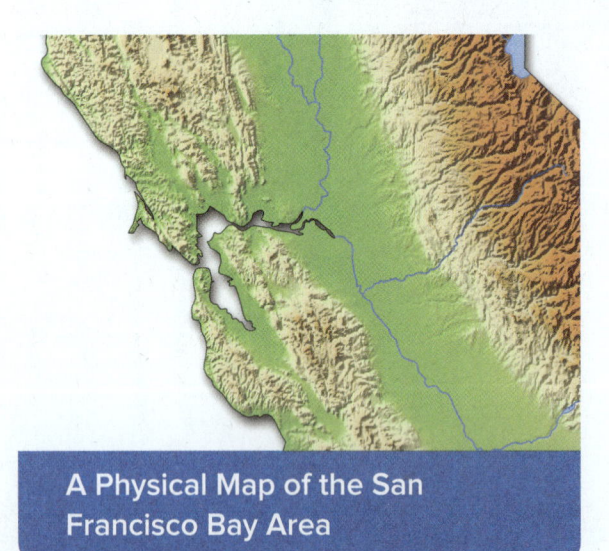

Quick Code: us4555s

Introducing Topographical Maps (Video)

A Physical Map of the San Francisco Bay Area

Major California Landforms
- Southern Cascade Mountains
- Great Basin
- Sierra Nevada
- Great Basin
- Coastal Ranges
- Salton Trough
- Great Central Valley
- Mojave Desert
- Peninsular Ranges

CCC Patterns

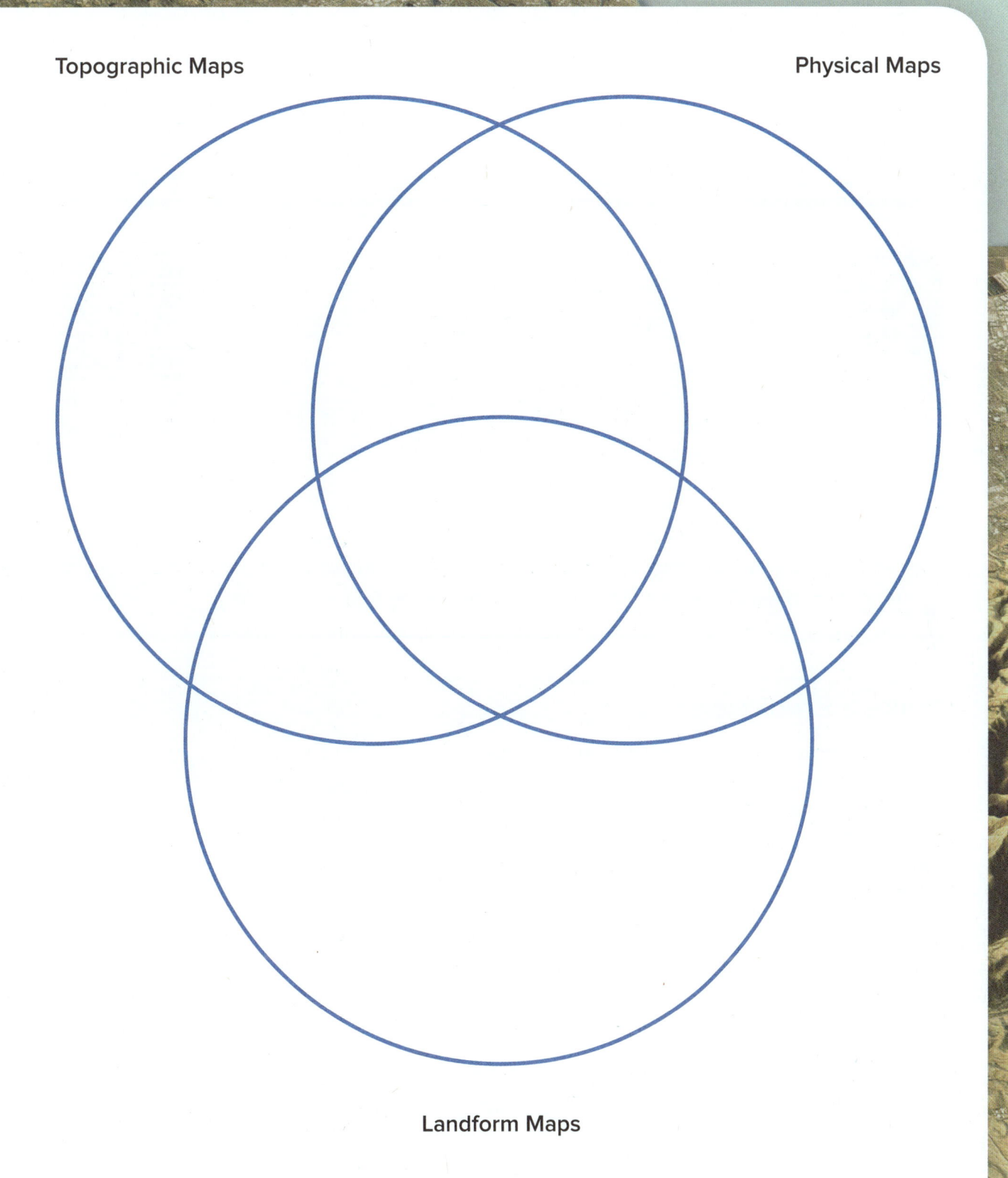

Topographic Maps

Physical Maps

Landform Maps

Concept 3.3: Mapping Landforms | 121

3.3 | Learn How can maps be used to discover patterns in Earth's landscape features?

What Are Some of the Major Landforms on Earth?

Activity 8

Investigate Like a Scientist

Quick Code: us4556s

Hands-On Investigation: Major Landforms on Earth

In this investigation, you will work in groups to analyze this topographic map of the world and then compare this map to an outline map of the world. On the outline, you will locate ocean basins and mountain ranges, on land and under the ocean, and sketch these features on the outline map. Look for patterns while you complete the activity.

Make a Prediction

There are mountains and canyons at the bottom of the ocean, just like there are on land. How can we show these landforms on maps? **Write** or **draw** your ideas below.

CCC Patterns

What materials do you need? (per group)

- Reference color-coded topographic map of the world
- Outline map of the world
- Computer access
- Colored pencils
- Pencils

What Will You Do?

1. Sketch the mountain ranges and ocean basins with the colored pencils.
2. Mark a series of "X" symbols where the mountain ranges begin and end.
3. Do the same with "O" symbols for the ocean basins.
4. Then, sketch the basic features of the mountain ranges and ocean basins. Use a different color to indicate each type of feature.

Concept 3.3: Mapping Landforms | 123

3.3 | Learn How can maps be used to discover patterns in Earth's landscape features?

Think About the Activity

Comment on the shape of most of the mountain ranges (Rockies, Andes, Himalayas, Alps, Appalachians). **Describe** the patterns you observe.

____

Comment on the shapes of the major ocean features. **Describe** the patterns you observe.

____

How are the mountain ranges positioned in relation to the continental margins and the middle of the oceans? **Describe** the patterns you observe.

How Can We Identify Smaller Landforms Using Maps?

Activity 9
Analyze Like a Scientist

Quick Code: us4557s

Comparing Maps and Satellite Images

Read the text and **compare** the topographic map of the Grand Canyon with the satellite image of the same area. Then, **answer** the questions that follow.

Comparing Maps and Satellite Images

The Grand Canyon can look different when different types of maps are used to show it. Maps are models of places. Look at the image of the Grand Canyon. It was taken using a satellite. It shows the path of the Colorado River. Are there any other features you recognize? Where on the image are the hills steepest? Can you detect ridges?

Grand Canyon Satellite Image

CCC Scale, Proportion, and Quantity

126

Now, look at the topographic map. It also shows the Grand Canyon, but in this map, more of the features are labeled. Some are represented by lines. Human-made features, such as roads, are clearly marked. Water features look brown on the satellite image, but they are marked in blue on the map. The faint brown lines represent hills.

These are contours, lines that join places of equal height (above sea level). Some of them have numbers. The numbers tell the height in meters above sea level. Where the contours are close together, the hills are steep. Where they are farther apart, the hills have gentler slopes. Maps usually come with a key or legend that explains the symbols used on them, to help you interpret them.

Topographic Map of Grand Canyon

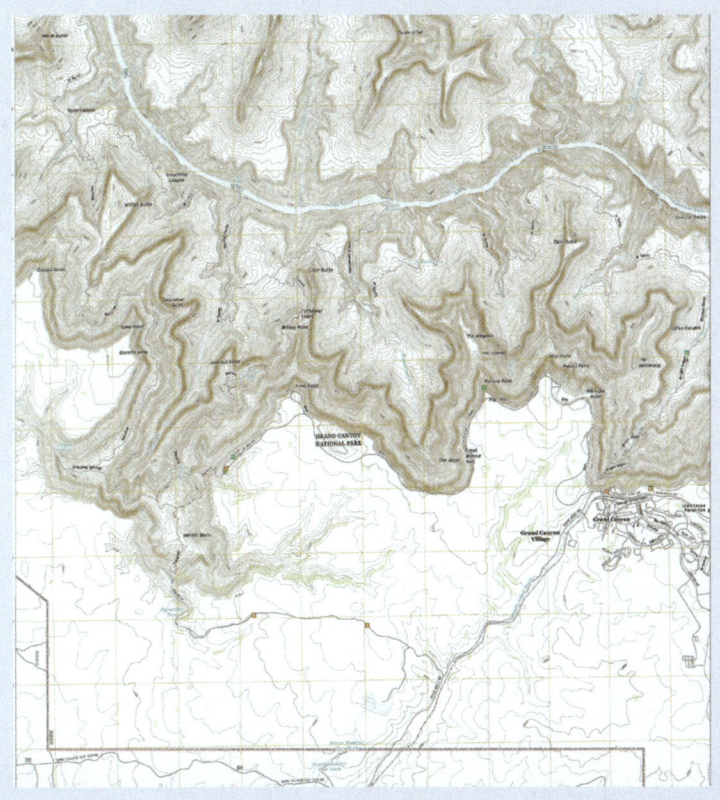

Concept 3.3: Mapping Landforms | 127

What do the contour lines represent?

How is the topographic map able to convey more information than the satellite image?

How Can We Use Maps to Tell Us More About Landforms?

Activity 10
Analyze Like a Scientist

Quick Code: us4558s

Watersheds

Read the text and **look** at the image. Then, **answer** the questions that follow.

Watersheds

Maps can be used to show the shape of landforms and the relationships between them. Streams flow down valleys and join to become rivers. Streams and rivers that are linked together and drain a common area called watersheds. A watershed is an area of land that feeds water resources. If all the water that falls on an area of land ends up in the same place, that area of land is called a watershed.

CCC Patterns

Concept 3.3: Mapping Landforms

Watersheds cont'd

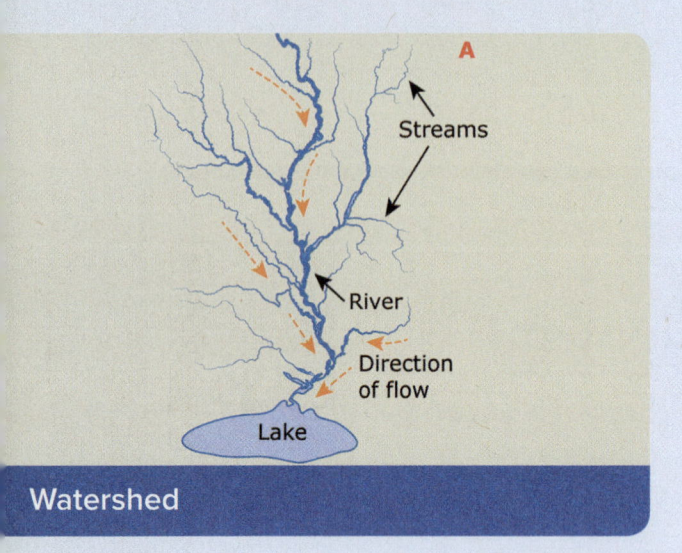

A
Streams
River
Direction of flow
Lake

Watershed

Think of a drop of rain falling on a ridgetop. It will either flow to one side of the ridge or to the other. The ridge separates two watersheds. Ridges are sometimes called divides. The Continental Divide, which runs north and south through the Americas, is an example. Water landing on the Continental Divide in North America will flow either toward the Pacific Ocean or toward the Gulf of Mexico.

What is a watershed?
_____
_____

What features on a map would you look for to identify a watershed?
_____
_____

130

Activity 11
Investigate Like a Scientist

Quick Code: us4559s

Hands-On Investigation: Mapping Rivers and Watersheds

In this investigation, you will explore how water flows and how changes to part of a watershed affect the region. You will work in groups to trace the movement of water as it flows, then locate a watershed in your region on a map.

Make a Prediction

First, **write** your predictions in the chart.

Question	My Prediction
In a watershed, what patterns would you observe about where the water originates and where it ends up?	
What water features could you identify on your watershed map? What patterns could you identify about the size of tributaries and the direction in which water flows?	
How can people use information about water sources and the flow of water in a watershed?	

SEP Analyzing and Interpreting Data

Concept 3.3: Mapping Landforms

3.3 | Learn How can maps be used to discover patterns in Earth's landscape features?

What materials do you need? (per group)

- Watershed map of your region or a nearby region
- Colored pencils

HANDS-ON INVESTIGATION

What Will You Do?

1. Look at the reference map online. Identify the water features.
2. Then, trace the direction of water flow.
3. Now, locate and trace the flow of water through a local watershed on a map of your region.
4. Draw arrows with the color pencils to trace the water flow.

Think About the Activity

Now, **examine** your map. **Think** about some critiques and **write** them in the T-Chart, together with possible ways to improve it.

Critiques	Suggested Improvements

3.3 | Learn — How can maps be used to discover patterns in Earth's landscape features?

Activity 12
Evaluate Like a Scientist

Quick Code: us4560s

Mapping Elevation, Mapping Elevation Continued, and Which Is Lower?

Mapping Elevation

Maps do not only use contour lines to provide elevation information. They also use colors to show elevation. **Study** the map, and then **circle** the correct answer to the question.

Physical Map of the United States and Canada

SEP Analyzing and Interpreting Data
CCC Scale, Proportion, and Quantity

134

Which landform has the highest elevation?

A. Appalachian Mountains

B. Baffin Island

C. Coastal Plain

D. Colorado Plateau

3.3 | Learn   How can maps be used to discover patterns in Earth's landscape features?

Mapping Elevation Continued

Look at the map, and **circle** the correct answer to the question.

Physical Map of the United States and Canada

What is the approximate elevation of the Great Plains?

A. 305 meters

B. 610 meters

C. 1,524 meters

D. 3,050 meters

Which Is Lower?

Look at the map, and **circle** the correct answer to the question.

Physical Map of the United States and Canada

Which is lower in elevation, the Coastal Plains of the eastern United States or the Great Plains? Why is this so?

A. Coastal Plains, because of faster erosion by the Great Lakes

B. Coastal Plains, because of faster erosion by the Atlantic Ocean

C. Great Plains, likely because of faster erosion by the Missouri River

D. Great Plains, likely because of faster erosion by the Rocky Mountains

Concept 3.3: Mapping Landforms | 137

3.3 | Learn How can maps be used to discover patterns in Earth's landscape features?

Activity 13
Observe Like a Scientist

Chesapeake Bay Watershed

As you **view** the image Chesapeake Bay Watershed, **look** for Earth's features in the image. Then, use the information in the image and what you have learned about mapping landforms to **answer** the questions.

Quick Code: us4561s

Chesapeake Bay Watershed

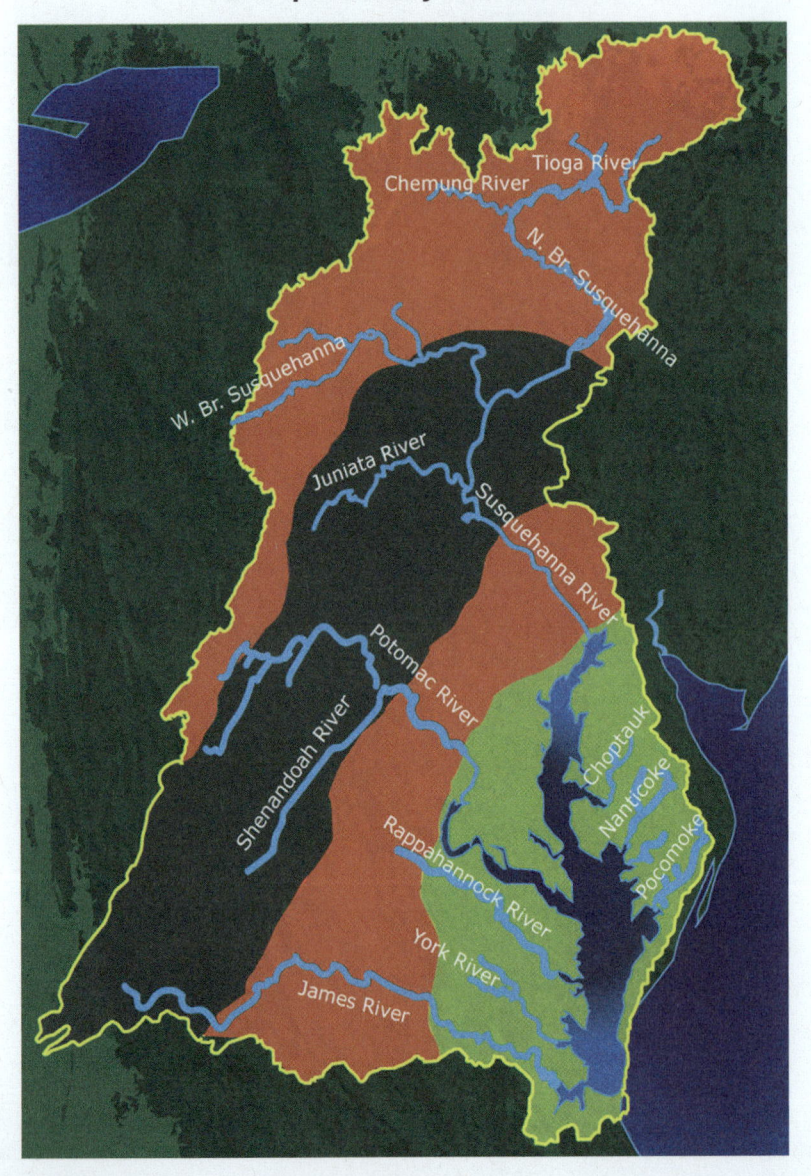

138

How would this watershed be affected if a change occurred near one of the tributaries? **Write** how each of these changes would affect the watershed. Then, **discuss** your answers with your class.

A factory is built on a river.

_____
_____
_____
_____
_____
_____
_____

A dam is built on a river.

_____
_____
_____
_____
_____
_____
_____

Concept 3.3: Mapping Landforms | 139

3.3 | Share — How can maps be used to discover patterns in Earth's landscape features?

Activity 14
Record Evidence Like a Scientist

Satellite Image of an Ocean

Now that you have learned about mapping landforms, look again at the satellite image of an ocean. You first saw this in Wonder.

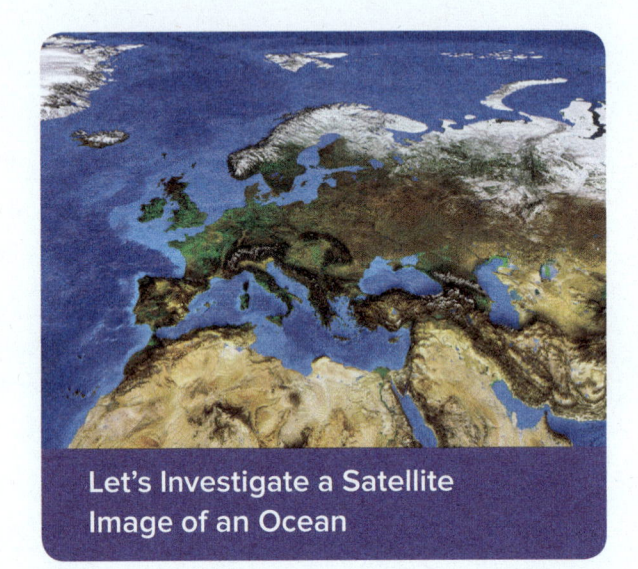

Quick Code: us4562s

Let's Investigate a Satellite Image of an Ocean

Talk Together

How can you describe the satellite image of an ocean now?

How is your explanation different from before?

SEP Constructing Explanations and Designing Solutions

Look at the Can You Explain? question. You first read this question at the beginning of the lesson.

> ### Can You Explain?
> How can maps be used to discover patterns in Earth's landscape features?

Now, you will use your new ideas about satellite images of an ocean to answer a question.

1. Choose a question. You can use the Can You Explain? question or one of your own. You can also use one of the questions that you wrote at the beginning of the lesson.

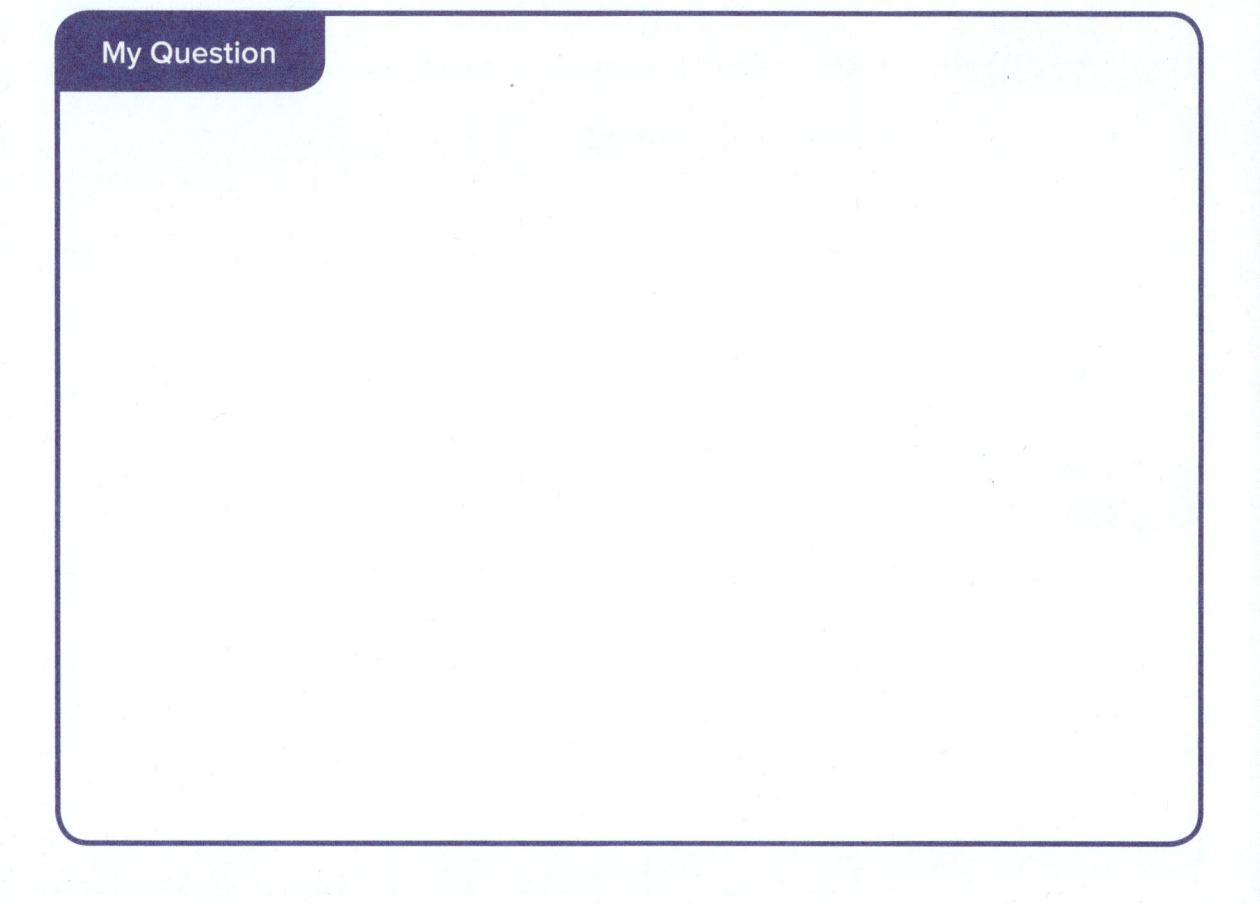

My Question

Concept 3.3: Mapping Landforms | 141

3.3 | Share
How can maps be used to discover patterns in Earth's landscape features?

2. Then, use the graphic organizers on the next pages to help you answer the question.

To plan your scientific explanation, first **write** your claim.

_____
_____
_____
_____

Next, **look** at your notes and maps from the activities Major Landforms on Earth and Mapping Rivers and Watersheds. Identify two pieces of evidence that support your claim:

Evidence 1

Evidence 2

Now, **write** your scientific explanation.

STEM in Action

Quick Code: us4563s

Activity 15
Analyze Like a Scientist

Ahoy, Captain!

Read the text and **watch** the videos. Then, **complete** the activities that follow.

Ahoy, Captain!

How do sailors know the way to go on the big ocean with only sea in every direction? Ancient navigators used the stars and sun to tell them the way. Some even used water temperature, currents, wind, sea animals, and birds to help them direct their boats. Pirates and famous captains who explored the world had compasses and made simple maps. Captains today have many tools to help them guide their ships through the waters of the world.

Early Sea Explorers (Video)

Captains use maps that show continents, islands, and ocean depth to help guide their ships. Ocean depth is extremely important for captains who are guiding ships. They don't want to get stuck or damage their ships. Many ships have sunk because the captain did not know the exact location of land, especially on stormy days. Sonar is one tool used to construct maps that show ocean depth to captains.

Sonar works by sending sound waves into the ocean. The sound waves are measured by how quickly they are reflected off of underwater objects. These measurements allow captains to know the exact depth of the water. Today, a captain will likely still refer to traditional paper maps, but will rely upon the sonar for real-time information about the location of the ship.

Ocean Exploration (Video)

Concept 3.3: Mapping Landforms

Sailing around the World

Can you imagine what it would be like to sail around the world on a ship?

Imagine that you are the captain of your own ship. You are going to **use** the map below and **find** an ocean route around the world. For the beginning of the trip, **choose** a large river that enters the ocean. **Label** all five oceans on your map and the continents. Then **make** a path around the world, and **choose** where at least three major rivers enter the ocean as stopping places. **Label** the rivers that you stop at and the city that is on the coast at the mouth of each of the rivers.

World Map

Describe your journey in your ship. **Include** your starting point and the names of the oceans and continents you pass and the general direction you are traveling. **Include** your stops on three major rivers with the names of the cities that are at the mouth of each river. **Conduct** additional research as needed to make sure your geography is accurate.

_____

_____

_____

_____

_____

_____

_____

_____

_____

_____

_____

_____

_____

_____

_____

_____

_____

Concept 3.3: Mapping Landforms

_____
_____
_____
_____
_____
_____
_____

Benefits of Sonar

Now that you have described your trip and the oceans you visited, consider how technology such as sonar could improve your voyage. **Read** all of the following statements, and **underline** the ones that correctly describe how sonar could affect a sail around the world.

- The sonar could help you know when there are hidden underwater obstacles that could harm the ship.

- The sonar could help you know how far it is to the next port.

- The sonar could help you know if a bay is deep enough for your ship to enter and anchor.

- The sonar could help you decide if the weather in the future will be good sailing weather.

- The sonar could help you locate food in the form of passing schools of fish.

Activity 16
Evaluate Like a Scientist

Quick Code: us4564s

Review: Mapping Landforms

Think about what you have read and seen. What did you learn?

Write down some core ideas you have learned. **Review** your notes with a partner. Your teacher may also have you take a practice test.

_____
_____
_____
_____
_____
_____
_____
_____

Talk Together

Think about what you saw in Get Started. Use your new ideas about changing landscapes to discuss the formation of the Grand Canyon.

SEP Obtaining, Evaluating, and Communicating Information

Concept 3.3: Mapping Landforms

CONCEPT 3.4

Volcanoes

Student Objectives

By the end of this lesson:

- ☐ I can explain, based on plate tectonics, the presence of volcanoes on all seven continents and under water.

- ☐ I can describe the processes that occur during volcanic eruptions and the factors that affect the types of volcanoes.

- ☐ I can represent data about Alaskan volcanoes in a bar graph to reveal patterns that indicate relationships among types of volcanoes.

- ☐ I can obtain and communicate information about the data sources and technologies scientists use to predict volcanic eruptions.

Key Vocabulary

- ☐ erupt
- ☐ lava
- ☐ magma
- ☐ tectonic plate

Quick Code: us4566s

Concept 3.4: Volcanoes | 151

Activity 1

Can You Explain?

What precautions can people take to plan for changes to the landscape?

Quick Code: us4567s

Concept 3.4: Volcanoes | 153

3.4 | Wonder
What precautions can people take to plan for changes to the landscape?

Activity 2
Ask Questions Like a Scientist

Quick Code: us4568s

Landscape Safety Check

Look at the sign. What do you think this sign means?

Let's Investigate Falling Rock Sign

Watch the video about Mount Vesuvius.

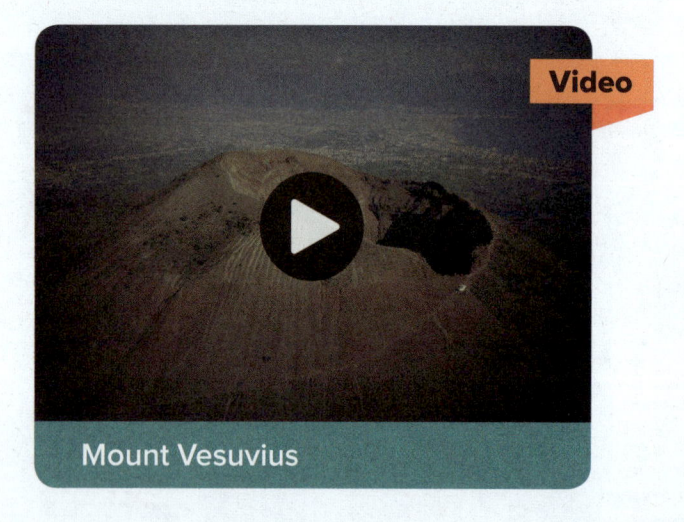

Mount Vesuvius

SEP Asking Questions and Defining Problems

154

Other landscape changes can happen quickly. **Look** at the images. What do you think caused these landscape changes?

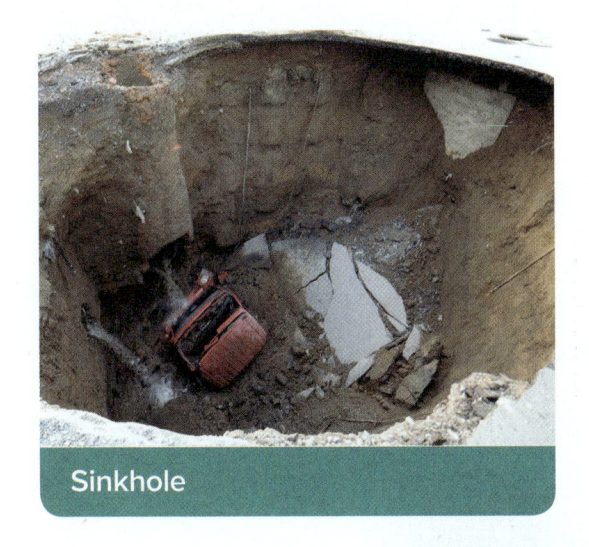

Sinkhole

Mudslide

Flash Flood

What do you wonder about volcanic eruptions and other landscape changes? **Write** 3 questions you have and **share** them with the class. Begin your questions with *What*, *Why*, and *When*.

_____

_____

_____

Concept 3.4: Volcanoes | 155

3.4 | Wonder
What precautions can people take to plan for changes to the landscape?

Activity 3
Observe Like a Scientist

Living on the Edge of Danger

As you **watch** the video, **look** for how volcanoes affect those people living near them.

Quick Code: us4569s

Living on the Edge of Danger (Video)

Talk Together

Now, talk together about why people would choose to live near a volcano.

Activity 4
Analyze Like a Scientist

Quick Code: us4570s

Pompeii

First, **predict** why people live near volcanoes. **Write** your prediction in the chart. Then, **read** the text. After you read, **write** whether your prediction was accurate in the chart.

Living Near Volcanoes	
Prediction	Observations

CCC Stability and Change

Concept 3.4: Volcanoes | 157

Pompeii

Pompeii was an ancient Roman city. It sat at the foot of a volcanic mountain that continuously smoked and rumbled. Around the city were fertile farmlands. In the year 79 CE, the volcano Mount Vesuvius erupted. The eruption destroyed the nearby city of Pompeii. In a very short time, the city was covered with 9 feet of ash and other volcanic material. Hundreds of years later, archaeologists unearthed the buried city. They discovered an eerie scene. People, pets, objects, and homes were preserved like statues in the ash, frozen in time. Why were so many people caught off guard by this eruption?

Mount Vesuvius is active and can still **erupt** at any time, yet many people still live in cities and towns at its base. People grow food in the healthy soils on its slopes, even as they know that the volcano poses a risk. People accept the risk of occasional eruptions to live in the fertile region. In fact, people live on many active volcanoes around the world.

Why are some mountains volcanic while others are not? What makes volcanoes erupt suddenly? What happens during volcanic eruptions? How can scientists predict eruptions and give people warning to get clear of danger? You will discover the answers to these questions as you investigate mountains and the nature of volcanoes in this lesson.

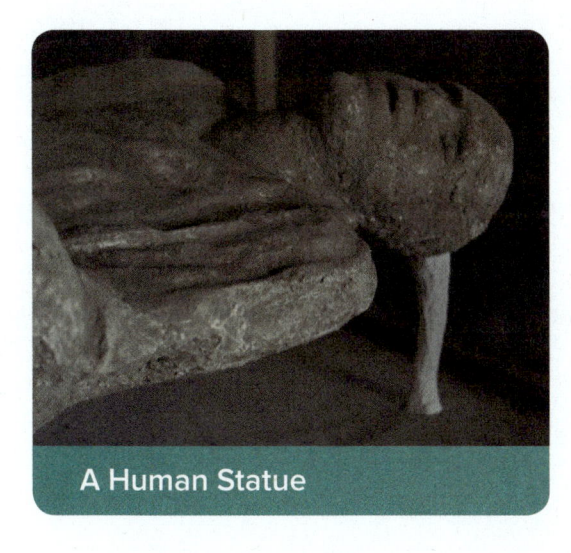

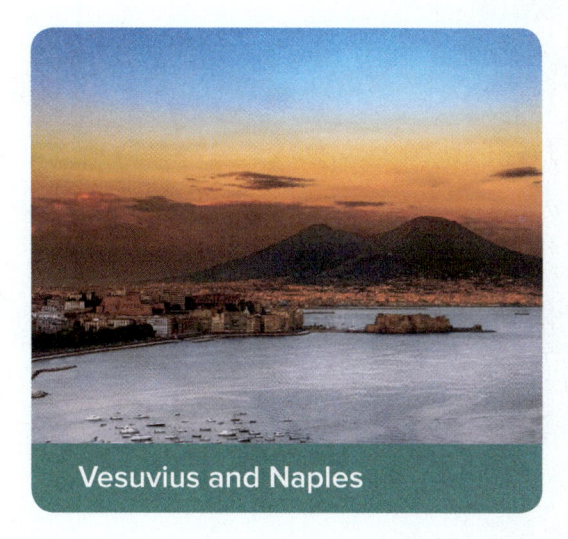

Now, **examine** the images of Mount Vesuvius and its location near the city of Naples. Then, **answer** the question that follows.

A Human Statue

Vesuvius and Naples

How can the people who live at the foot of Mount Vesuvius protect themselves from volcanic eruption?

_____

_____

_____

_____

_____

Concept 3.4: Volcanoes

3.4 | Wonder
What precautions can people take to plan for changes to the landscape?

Activity 5
Evaluate Like a Scientist

Quick Code: us4571s

What Do You Already Know About Volcanoes?

Volcanic Materials

Which materials could come out of a volcano during an eruption? **Circle** all that apply.

- A. lava
- B. gases
- C. meteorite
- D. ash

CCC Energy and Matter

Volcanic Activity

Consider what you know about volcanic activity. **Underline** the TWO statements that are true.

When volcanoes erupt, they always feature lava and ash shooting high into the air.

All volcanoes formed in the distant past, but there are some old volcanoes that are still active today.

Most volcanoes formed in the distant past, but new active volcanoes form even today.

All volcanoes always begin forming in the oceans.

Volcanic eruptions can differ in their size and severity; some slowly ooze lava, and others erupt violently.

3.4 | Learn
What precautions can people take to plan for changes to the landscape?

Where Are Volcanoes Found on Earth?

Activity 6
Observe Like a Scientist

Volcano Locations

As you **view** the images, **look** for volcano locations. Then, **answer** the questions that follow.

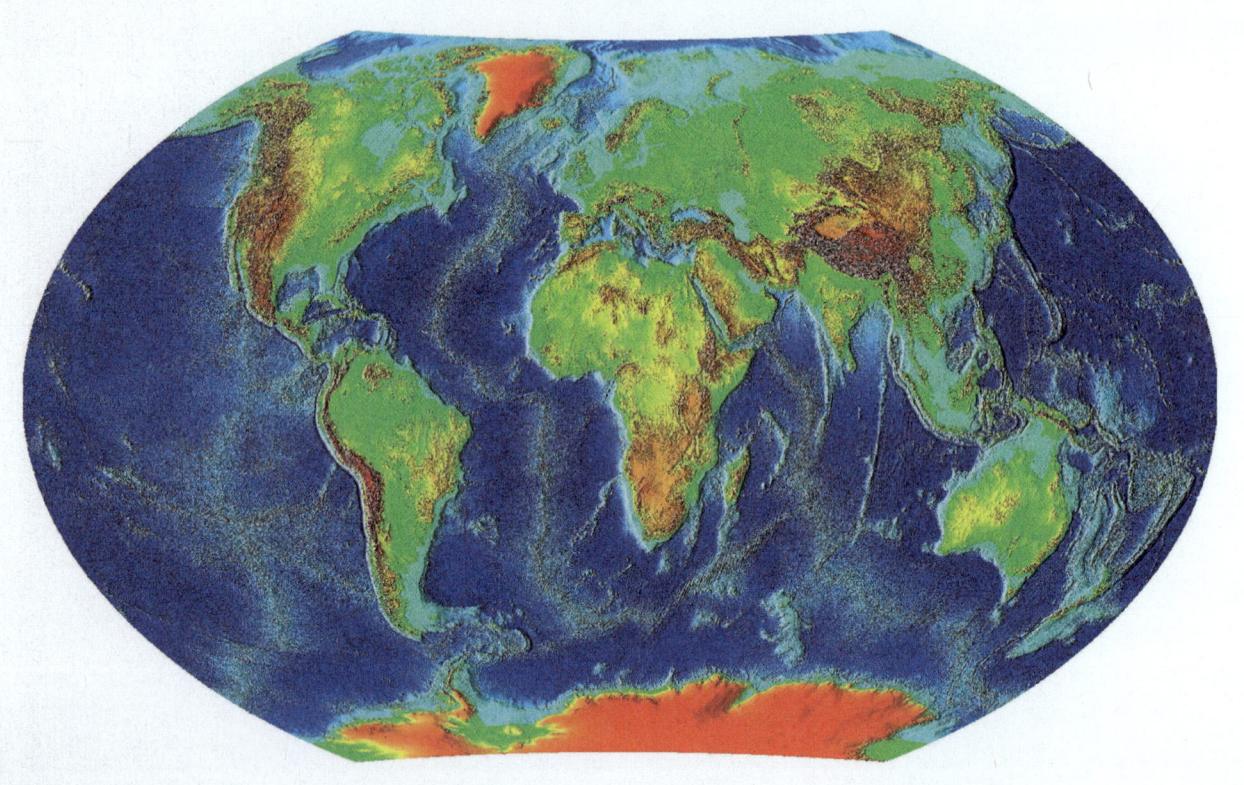

Quick Code: us4572s

Topographic Map of the World

`CCC` Patterns

162

Volcanoes on Earth

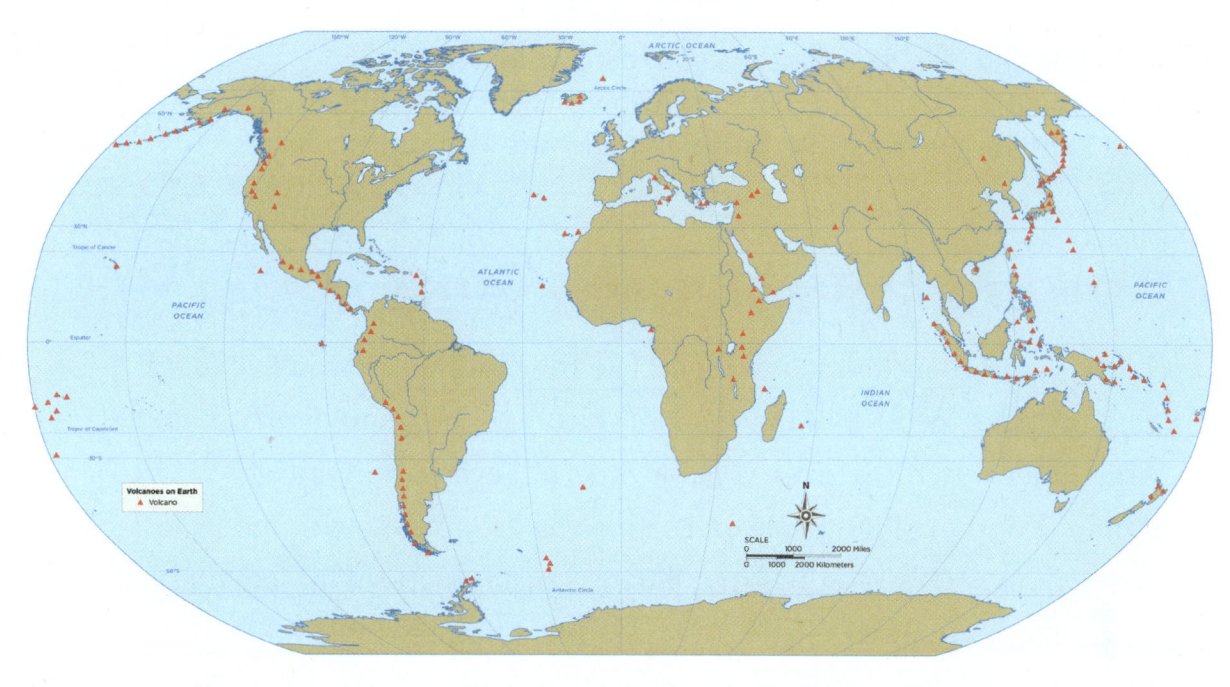

What relationships or patterns are found between the two images?

_____

_____

_____

What conclusion about volcano locations on Earth can you make?

_____

_____

_____

Concept 3.4: Volcanoes | 163

Activity 7
Analyze Like a Scientist

Quick Code: us4573s

Volcanoes and Earth's Crust

Read the text and view the image. As you read, **write** questions you have about the text in the chart that follows. Then, work with a partner to **answer** the questions. **Write** your answers in the chart.

Volcanoes and Earth's Crust

Earth's surface includes land areas, like the place where you live. But it also includes the floor of the oceans. Volcanoes are found both on land and under water. There are volcanoes on all seven continents, even on Antarctica, the coldest place on Earth. Why are volcanoes found in so many places? One reason is that Earth's crust is made of large plates. These **tectonic plates** move slowly in different directions. When these plates run into each other or move apart, volcanoes form. When plates move together, mountains are pushed up.

Volcanoes and Plate Boundaries

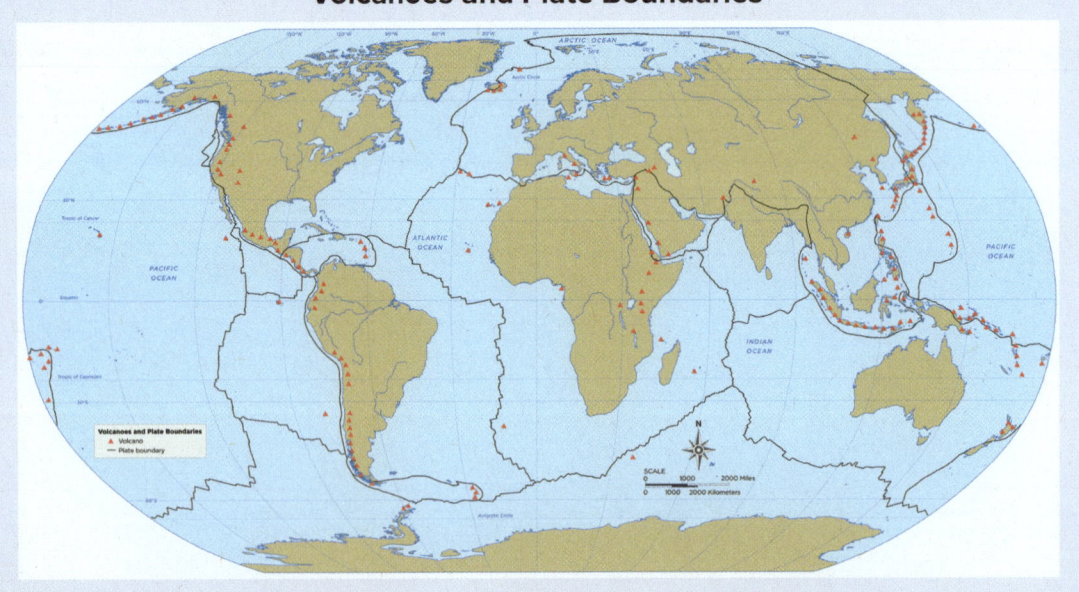

CCC Patterns

Questions	Answers

Concept 3.4: Volcanoes | 165

3.4 | Learn
What precautions can people take to plan for changes to the landscape?

How Are Volcanoes Formed?

Activity 8
Observe Like a Scientist

Quick Code: us4574s

How Are Volcanoes Formed?

Look at the images. Look for volcano shapes and features. Then, **write** questions about the possible causes and effects of volcanoes using the chart that follows.

- Composite Volcano
- Stromboli Volcano
- Erupting Volcano
- Lava Channel

CCC Cause and Effect

Cause	Event	Effect

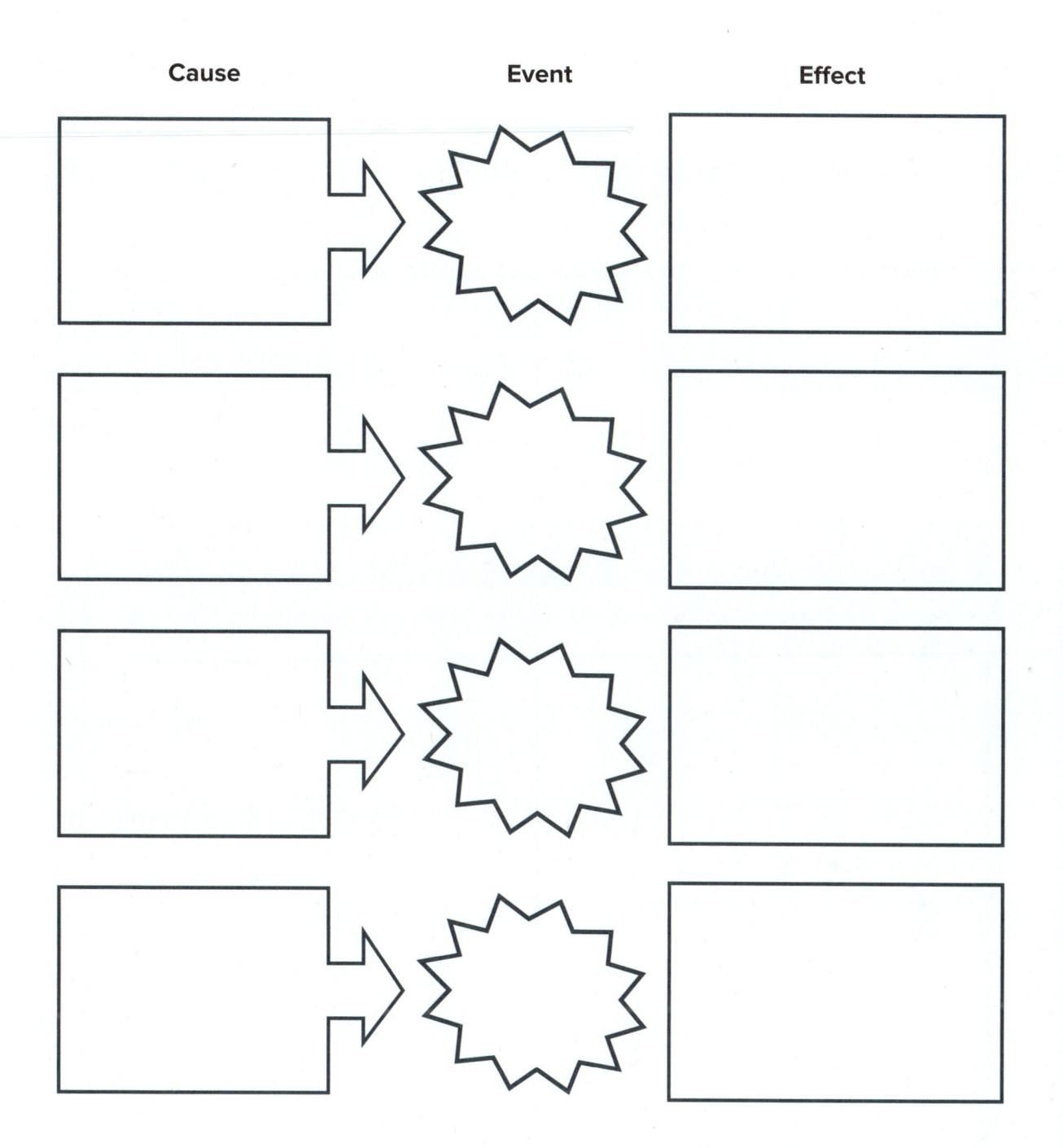

Concept 3.4: Volcanoes

3.4 | Learn
What precautions can people take to plan for changes to the landscape?

Activity 9
Investigate Like a Scientist

Hands-On Investigation: Cake Batter Lava

Quick Code: us4575s

In this investigation, you will use cake batter to model lava flows. Remember that a volcanologist is someone who studies volcanoes. You will map the lava flows like volcanologists do.

Make a Prediction

First, **write** your predictions in the chart.

Questions	My Prediction
How will the steepness of a volcano's sides affect how lava flows from it?	
How do you think making maps of lava flows helps volcanologists protect people?	

SEP Developing and Using Models
CCC Patterns

What materials do you need? (per group)

- Container of cake batter (mixed prior to class by teacher)
- Aluminum foil pan, 13 × 9 × 2
- Books about one inch thick each, 6
- Plastic cup, 9 oz
- Paper clips, large
- Clear acetate sheet
- Pencils
- Metric ruler
- Stopwatch
- Graph paper
- Computer drawing program (optional)

What Will You Do?

1. Pour the batter on the higher end of the foil pan.
2. Record the time and distance of the flow based on the marks on the paper sheet.
3. Insert an unfolded paper clip into the batter to measure the depth of the flow.
4. Use a metric ruler to measure the width of the flow.
5. Record your observations in the data table.
6. Draw a map of how the flow spreads on the grid.
7. Then, raise the end of the foil pan to increase the slope.
8. Repeat the previous steps under this new condition.
9. Record any change you observe.

Concept 3.4: Volcanoes

3.4 | Learn
What precautions can people take to plan for changes to the landscape?

	Height of Baking Sheet	Time	Distance of Flow	Width of Flow	Depth of Flow
Trial 1					
Trial 2					

Think About the Activity

Now, **examine** your model. **Identify** its limitations and **write** them in the T-Chart, together with possible solutions.

Limitations	Possible Solutions

Concept 3.4: Volcanoes

Activity 10
Analyze Like a Scientist

Quick Code: us4576s

Studying Volcanoes—All in a Day's Work

As you **read** this text, **underline** parts of the text using the following code.

___ - Underline the main idea

_ _ _ - Dotted underline the supporting details

? - To indicate the parts you don't understand

! - To indicate interesting or exciting facts

Studying Volcanoes—All in a Day's Work

Volcanoes have always fascinated people. What makes them explode with hot liquid? How fast does that liquid travel? How do you know when one will erupt? These questions and more may be answered by a volcanologist.

Volcanologists study or teach about volcanoes. A volcano is a mountain or an opening in Earth's crust where **magma** reaches Earth's surface. Magma is hot liquid rock. When pressure from the magma builds up inside a volcano, it may erupt. **Lava**, or magma that reaches Earth's surface, can shoot out. Gases, ash, and rock can produce a dark cloud above the volcano.

CCC Stability and Change

Volcanoes can be grouped based on how likely they are to erupt. Some volcanoes are active. They have recently erupted, or they may erupt soon. Volcanologists watch active volcanoes very closely. Some volcanoes are dormant, or sleeping. They have not erupted for a long time. Dormant volcanoes show signs that they may erupt again. Some volcanoes are extinct. They haven't erupted for thousands of years.

What kinds of things do you think volcanologists study? Many volcanologists work in the field collecting data from volcano sites. Some try to predict eruptions. They study things like movements within Earth's crust that may start an eruption. Some volcanologists study lava flows. They use a process called mapping. Special tools tell them how fast lava is moving toward areas where people live.

No matter what kind of work they do, volcanologists stay very busy. After all, there are more than 1,500 active volcanoes in the world!

This volcano in Japan is active.

Concept 3.4: Volcanoes

3.4 | Learn
What precautions can people take to plan for changes to the landscape?

Activity 11
Observe Like a Scientist

Types of Volcanoes

As you **watch** the video, **look** for landforms, volcano locations, and changes in Earth's surface.

Quick Code: us4577s

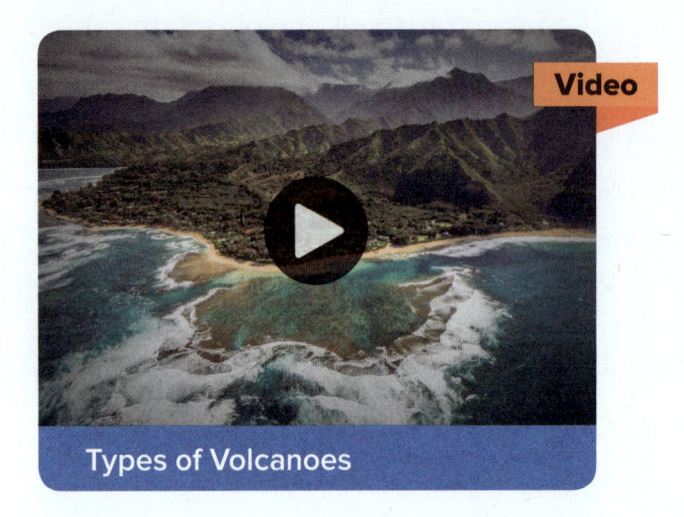

Video

Types of Volcanoes

Talk Together

Now, talk together about how Earth's interior helps shape its exterior.

CCC Patterns

Activity 12
Observe Like a Scientist

When Earth Erupts

As you **watch** the video, **look** for different types of volcanoes.

Quick Code: us4578s

When Earth Erupts

Talk Together

Now, talk together about how the three types of volcanoes are similar and different.

CCC Patterns

Concept 3.4: Volcanoes | 175

Activity 13
Analyze Like a Scientist

Quick Code: us4579s

Out of the Volcano!

Read the text and **view** the image. Then, **answer** the questions that follow.

Out of the Volcano!

In the past, volcanoes erupted much more frequently than they do today. We are lucky that most volcanoes have become very quiet. When a volcano does erupt, it can cause great damage very quickly.

Many volcanoes emit lava when they erupt. Lava is a mixture of melted rocks and minerals that flows over the land. Many gases are mixed into the lava, too.

Lava is incredibly hot. When it flows, it burns any living things in its path. You might think that boiling water is hot. Yet, while water boils at 100°C, the temperature of lava can be 1,000°C or higher! Fortunately, lava cools very quickly in the air, and even more quickly in water.

Not all lava is alike. Different lava mixtures contain different combinations of minerals. Some mixtures flow quickly, and they may travel many meters away from the volcano's opening before hardening into rock. Other mixtures flow more slowly. They harden near the

SEP Developing and Using Models
CCC Patterns

volcano's opening. Over many years, the hardened rocks often build up into the cone shape that is common among volcanoes. Still other mixtures hardly flow at all. They might harden while still inside or beneath a volcano. It all depends on the substances contained in the lava.

Lava is not a single substance, but instead is a mixture of rocks, minerals, and gases. It flows like a liquid, but the rocks harden as they meet the cool air.

Concept 3.4: Volcanoes | 177

How does cake batter compare and contrast to lava flowing down a mountain?

_____
_____
_____
_____
_____
_____
_____
_____

If scientists would use a cake batter model to represent lava flow to the general public, what types of dangers should they point out that the model does not represent well?

_____
_____
_____
_____
_____
_____
_____
_____
_____
_____

How Can Scientists Predict Volcanic Eruptions?

Activity 14

Evaluate Like a Scientist

Quick Code: us4580s

Alaskan Volcanoes

Read the text. Use this information from the U.S. Geological Survey to **make** a bar graph about Alaskan volcanoes.

Alaskan Volcanoes

The United States has about 150 active volcanoes. Alaska is home to about one-third of them. While all of these are expected to erupt again someday, few are currently erupting. Mount Pavlof has sent out a lava fountain and volcanic ash as high as 22,000 feet above sea level. Mount Cleveland has had several explosions with small lava flows and ash shooting up to 35,000 feet. About half of Alaska's active volcanoes are monitored just in case gas comes out or the ground changes around them.

Living with Volcanoes

SEP Analyzing and Interpreting Data

Concept 3.4: Volcanoes

3.4 | Learn
What precautions can people take to plan for changes to the landscape?

Now, **create** your graph. Your graph should show:

- active volcanoes
- dormant volcanoes
- volcanoes that are currently erupting

Activity 15
Analyze Like a Scientist

Quick Code: us4581s

Monitoring Volcanoes

Read the text and **view** the image. Then, **answer** the question that follows.

Monitoring Volcanoes

We cannot stop volcanoes from erupting. However, scientists who study volcanoes, called volcanologists, can predict whether or when they are likely to erupt. They can also predict what form the eruption may take. For example, a volcanologist uses equipment to investigate gases coming out of a volcano. Some technologies can be used to test volcanic materials on site.

Volcanic Monitoring

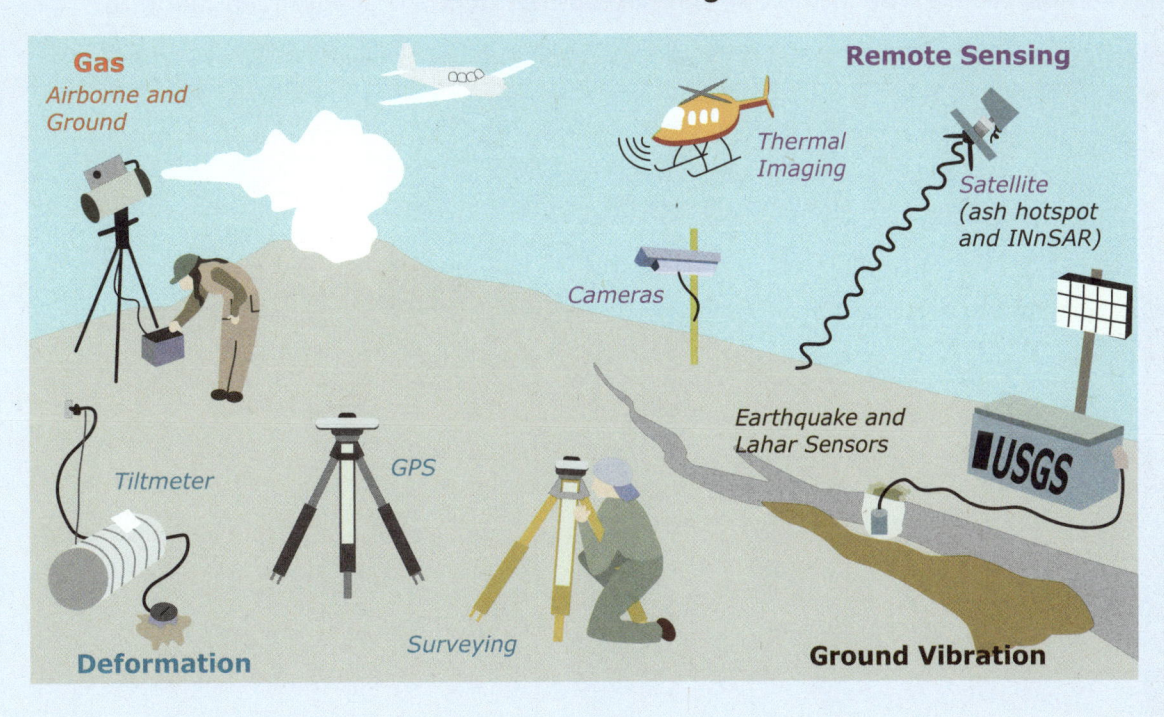

- **Gas** — Airborne and Ground
- Thermal Imaging
- **Remote Sensing** — Satellite (ash hotspot and INnSAR)
- Cameras
- Tiltmeter
- GPS
- Surveying
- Earthquake and Lahar Sensors (USGS)
- **Deformation**
- **Ground Vibration**

Concept 3.4: Volcanoes | 181

Monitoring Volcanoes *cont'd*

Other technologies can be used to study volcanoes from farther away. For example, tiltmeters measure the slope of the volcanoes and strainmeters measure whether the ground is being stretched.

Both of these changes could indicate a build-up of magma from below and a possible eruption. Using special thermometers to measure changes in temperature in active vents, scientists can monitor gases seeping from the ground. Seismometers can be used to detect earthquakes that indicate magma that is moving underground.

Why would it be important to have different types of volcanic monitoring?

Activity 16
Observe Like a Scientist

Predicting Volcanic Eruptions

As you **watch** the video, **look** for different techniques that scientists use in the field.

Quick Code: us4582s

Predicting Volcanic Eruptions (Video)

Talk Together

Now, talk together about the techniques scientists use to predict the size of an eruption. What risks do scientists face when they take these measurements?

CCC Patterns

Concept 3.4: Volcanoes

3.4 | Learn — What precautions can people take to plan for changes to the landscape?

Activity 17
Evaluate Like a Scientist

Quick Code: us4583s

Studying Volcanoes

Match each technology used to study volcanoes with the information it provides. **Draw** a line from the technology to the information type.

Technology
Seismometer
Thermometer
Tiltmeter

Information Type
Mapping hot magma in active vents inside the volcano
Measuring the expansion of a volcano
Detecting movement of magma inside the volcano

Concept 3.4: Volcanoes | 185

3.4 | Share
What precautions can people take to plan for changes to the landscape?

Activity 18
Record Evidence Like a Scientist

Quick Code: us4584s

Landscape Safety Check

Now that you have learned about volcanoes and other hazards, look again at the falling rock sign. You first saw this in Wonder.

Let's Investigate Landscape Safety Check

Talk Together

How can you describe the falling rock sign now?

How is your explanation different from before?

SEP Constructing Explanations and Designing Solutions

Look at the Can You Explain? question. You first read this question at the beginning of the lesson.

> **Can You Explain?**
>
> What precautions can people take to plan for changes to the landscape?

Now, you will use your new ideas about landscape safety to answer a question.

1. Choose a question. You can use the Can You Explain? question or one of your own. You can also use one of the questions that you wrote at the beginning of the lesson.

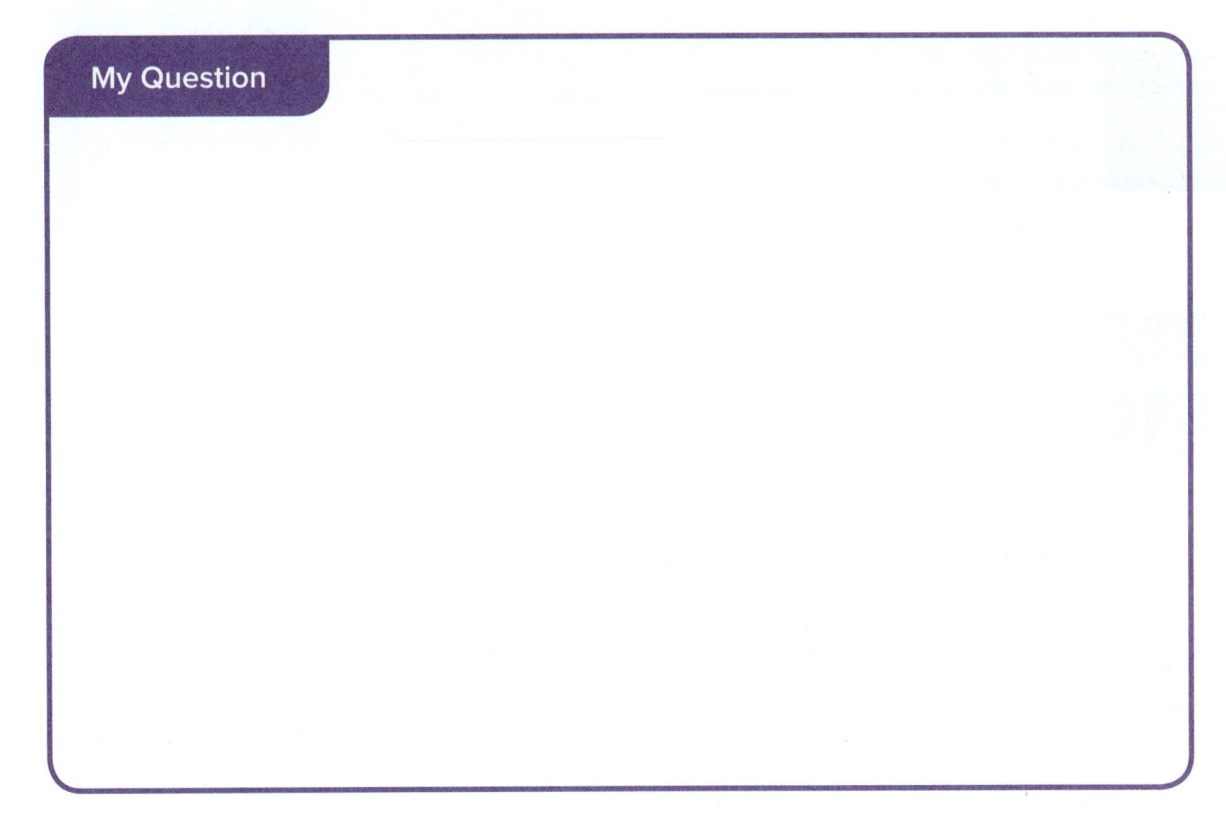

My Question

2. Then, use the graphic organizers on the next pages to help you answer the question.

Concept 3.4: Volcanoes

3.4 | Share
What precautions can people take to plan for changes to the landscape?

To plan your scientific explanation, first **write** your claim.

My claim:
_____
_____
_____
_____

Next, **look** at your notes and answers for the activities Cake Batter Lava, Lava Flows, and Monitoring Volcanoes. Identify two pieces of evidence that support your claim:

Evidence 1

Evidence 2

Now, **write** your scientific explanation.

Concept 3.4: Volcanoes | 189

STEM in Action

Quick Code: us4585s

Activity 19
Analyze Like a Scientist

Clues beneath Earth's Surface

Read the text. As you read, **highlight** the responsibilities of an igneous petrologist and **underline** the responsibilities of a volcanologist.

Clues beneath Earth's Surface

Studying volcanic rocks gives us important information about what is under Earth's surface. People do not have the technology to drill deep enough into Earth to see what is happening, but studying volcanic rocks gives scientists clues. Scientists can use these clues to learn about conditions deep below Earth's surface. A scientist who studies volcanic rocks is called an igneous petrologist.

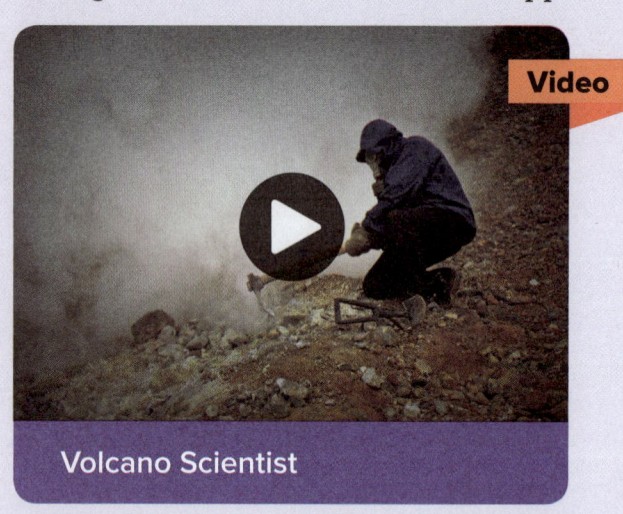

Video: Volcano Scientist

While igneous petrologists study cooled rocks from ancient volcanoes, volcanologists study currently active volcanoes. These scientists help give people who live near volcanoes important information about the volcanic activity. Volcanologists study gases that come out of volcanoes between eruptions. They study actively erupting volcanoes. They also look for clues in seismic activity. This involves measuring waves that pass through rock as magma starts to rise.

Volcanologists work with local officials in areas where people live near volcanoes. They work together to inform people about the risk of volcanic activity. Emergency workers help people leave the area. They can also help rescue people and clean up after an eruption.

Can you think of other careers that require knowledge of volcanoes?

Volcanology: The Science of Predicting Volcanoes

Concept 3.4: Volcanoes | 191

Volcano Scientists

Draw a line to match each type of volcano scientists on the left with the correct job goals and duties on the right. The scientists will be matched with more than one duty.

Type of Scientist
Volcanologist
Igneous Petrologist

Job Goals and Duties
Studies rocks formed in ancient volcanoes.
Collects samples from active volcanoes.
Learns about processes deep below Earth's surface.
Gains information that can be used to predict eruptions.

Concept 3.4: Volcanoes | 193

3.4 | Share — What precautions can people take to plan for changes to the landscape?

Activity 20
Evaluate Like a Scientist

Quick Code: us4586s

Review: Volcanoes

Think about what you have read and seen. What did you learn?

Write down some core ideas you have learned. **Review** your notes with a partner. Your teacher may also have you take a practice test.

_____
_____
_____
_____
_____
_____
_____
_____
_____
_____
_____
_____
_____
_____
_____

SEP Obtaining, Evaluating, and Communicating Information

Talk Together

Think about what you saw in Get Started. Use your new ideas about volcanoes to discuss the formation of the Grand Canyon.

Unit Project

Solve Problems Like a Scientist

Quick Code: us4588s

Unit Project: Lava Flows and the Grand Canyon

In this project, you will use what you know about Earth's surface changes to model how lava affected the Grand Canyon.

The Grand Canyon is formed partially by the erosion and weathering forces of the Colorado River. In addition, volcanic activity near the Grand Canyon has influenced its geography. Using special cameras, scientists can detect minerals indicative of lava flows. These lava flows spill down the walls of the canyon all the way to the river.

Look at the image below taken by NASA of the lava flows around the Grand Canyon. Think about how lava flows can affect the Grand Canyon and consider a model that can show this event. Then, complete the activity that follows.

Lava Flows and the Grand Canyon

SEP Developing and Using Models   **CCC** Cause and Effect

Prediction

Look at the picture of the Grand Canyon.

Grand Canyon Lava Flow

Predict what would happen if the lava flows down the canyon walls into the canyon.

How will this affect the path of the Colorado River?

_____
_____
_____
_____

What will happen over time?

_____
_____
_____
_____

Unit 3: Earth's Changing Surface

Unit Project

Model

Consider how you can use liquid glue to build a model to predict what will happen to the Grand Canyon. How will you represent the canyon? How will you represent the lava? **Draw** a picture of your design.

Build your model. Then, **summarize** your findings. **Use** the Summary Frames to show what happened.

_____   _____   _____

_____   _____   _____

Unit 3: Earth's Changing Surface | 199

Grade 4 Resources

- **Bubble Map**
- **Safety in the Science Classroom**
- **Vocabulary Flash Cards**
- **Glossary**
- **Index**

Name _____

Bubble Map

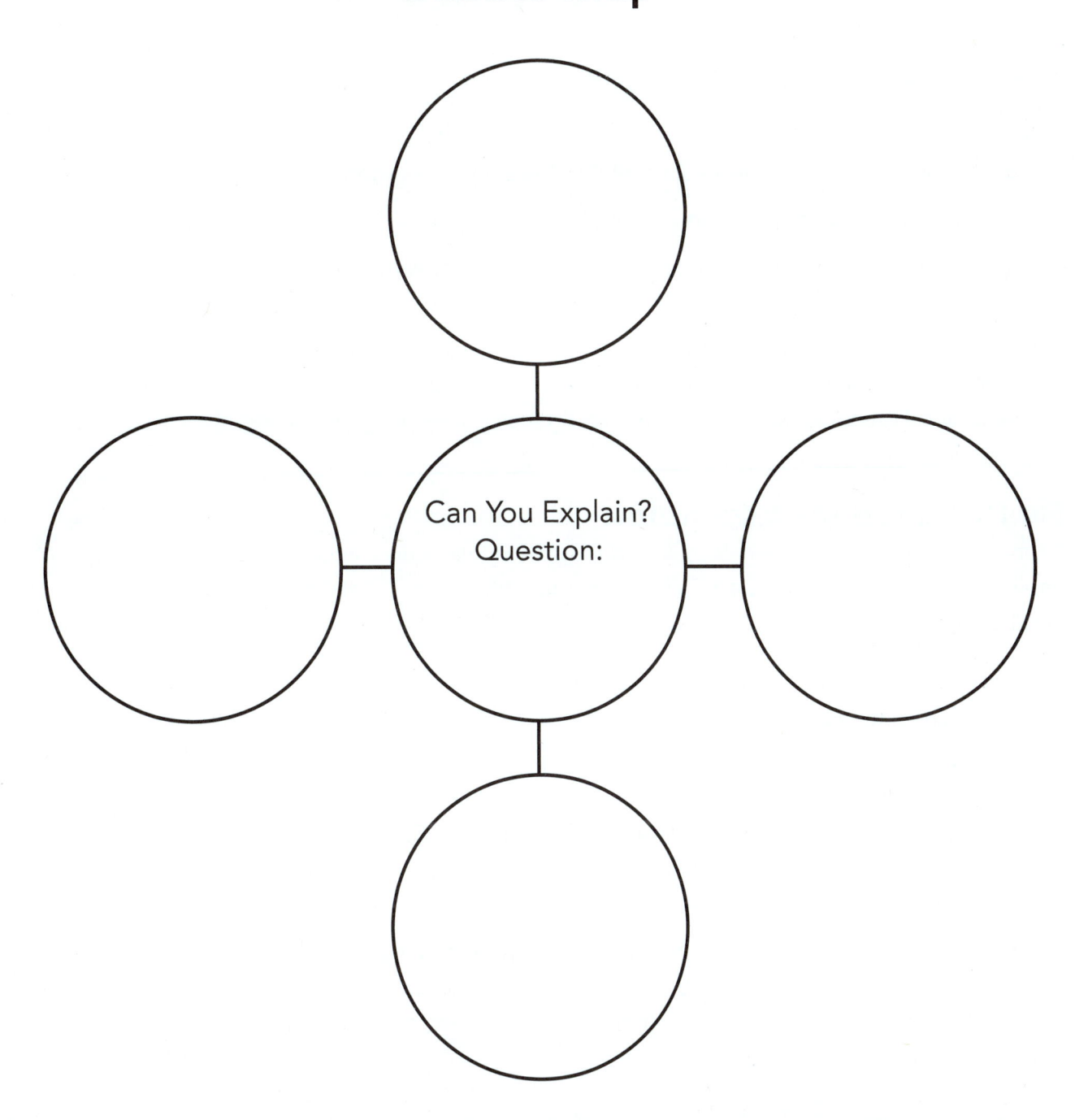

Can You Explain? Question:

Bubble Map | R3

Safety

Safety in the Science Classroom

Following common safety practices is the first rule of any laboratory or field scientific investigation.

Dress for Safety

One of the most important steps in a safe investigation is dressing appropriately.

- Splash goggles need to be kept on during the entire investigation.
- Use gloves to protect your hands when handling chemicals or organisms.
- Tie back long hair to prevent it from coming in contact with chemicals or a heat source.
- Wear proper clothing and clothing protection. Roll up long sleeves, and if they are available, wear a lab coat or apron over your clothes. Always wear close toed shoes. During field investigations, wear long pants and long sleeves.

Be Prepared for Accidents

Even if you are practicing safe behavior during an investigation, accidents can happen. Learn the emergency equipment location in your classroom and how to use it.

- The eye and face wash station can help if a harmful substance or foreign object gets into your eyes or onto your face.
- Fire blankets and fire extinguishers can be used to smother and put out fires in the laboratory. Talk to your teacher about fire safety in the lab. He or she may not want you to directly handle the fire blanket and fire extinguisher. However, you should still know where these items are in case the teacher asks you to retrieve them.
- Most importantly, when an accident occurs, immediately alert your teacher and classmates. Do not try to keep the accident a secret or respond to it by yourself. Your teacher and classmates can help you.

Practice Safe Behavior

There are many ways to stay safe during a scientific investigation. You should always use safe and appropriate behavior before, during, and after your investigation.

- Read the all of the steps of the procedure before beginning your investigation. Make sure you understand all the steps. Ask your teacher for help if you do not understand any part of the procedure.

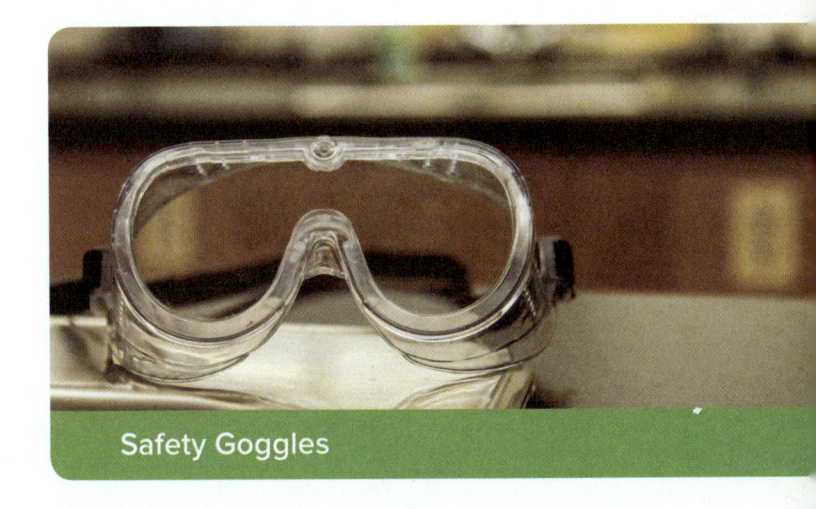

- Gather all your materials and keep your workstation neat and organized. Label any chemicals you are using.

- During the investigation, be sure to follow the steps of the procedure exactly. Use only directions and materials that have been approved by your teacher.

- Eating and drinking are not allowed during an investigation. If asked to observe the odor of a substance, do so using the correct procedure known as wafting, in which you cup your hand over the container holding the substance and gently wave enough air toward your face to make sense of the smell.

- When performing investigations, stay focused on the steps of the procedure and your behavior during the investigation. During investigations, there are many materials and equipment that can cause injuries.

- Treat animals and plants with respect during an investigation.

- After the investigation is over, appropriately dispose of any chemicals or other materials that you have used. Ask your teacher if you are unsure of how to dispose of anything.

- Make sure that you have returned any extra materials and pieces of equipment to the correct storage space.

- Leave your workstation clean and neat. Wash your hands thoroughly.

Safety Goggles

Safety in the Science Classroom

Vocabulary Flash Cards

air
the part of the atmosphere closest to Earth; the part of the atmosphere that organisms on Earth use for respiration

canyon
a deep valley carved by flowing water

chemical weathering
changes to rocks and minerals on Earth's surface that are caused by chemical reactions

delta
a fan-shaped mass of mud and other sediment that forms where a river enters a large body of water

Vocabulary Flash Cards | R7

deposition

laying sediment back down after erosion moves it around

dune

a hill of sand created by the wind

erosion

the removal of weathered rock material; the small particles are transported by wind, water, ice, and gravity

erupt

the action of lava coming out of a hole or crack in Earth's surface; the sudden release of hot gasses or lava built up inside a volcano

Vocabulary Flash Cards | R9

glacier

a large sheet of ice or snow that moves slowly over Earth's surface

heat

the transfer of thermal energy

landform

a large natural structure on the earth's surface such as a mountain, a plain, or a valley

lava

molten rock that comes through cracks in Earth's crust; liquid and gas mixture cools and turns into solid rock

Vocabulary Flash Cards | R11

magma

melted rock located beneath Earth's surface

map

a flat model of an area

meander

winding or indirect movement or course

mountain

an area of land that forms a peak at a high elevation

Vocabulary Flash Cards | R13

satellite

a natural or artificial object that revolves around another object in space

sediment

solid material, moved by wind and water, that settles on the surface of land or the bottom of a body of water

soil

the outer layer of Earth's crust; made of bits of dead plant and animal material, rocks, and minerals

tectonic plate

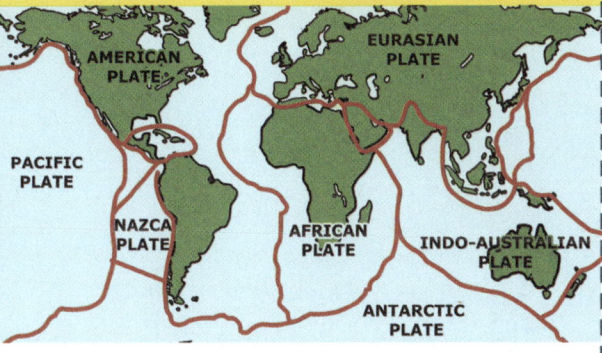

one of several huge pieces of Earth's crust

Vocabulary Flash Cards | R15

topographic map

a map which shows the relief and other features of an area

volcano

an opening in the Earth's surface through which magma and gases or only gases erupt

water

a compound made of hydrogen and oxygen

weathering

the physical or chemical breakdown of rocks and minerals into smaller pieces or aqueous solutions on Earth's surface

Vocabulary Flash Cards | R17

Glossary

English ——— A ——— Español

acceleration to increase speed	**aceleración** aumentar la rapidez
adaptation something a plant or animal does to help it survive in its environment (related word: adapt)	**adaptación** algo que hace una planta o un animal para sobrevivir en su medio ambiente (palabra relacionada: adaptar)
air the part of the atmosphere closest to Earth; the part of the atmosphere that organisms on Earth use for respiration	**aire** parte de la atmósfera más cercana a la Tierra; la parte de la atmósfera que los organismos que habitan la Tierra utilizan para respirar
amplitude height or "strength" of a wave	**amplitud** altura o "magnitud" de una onda
analog one continuous signal that does not have any breaks	**analógico** señal continua que no tiene ninguna interrupción

antenna
a device that receives radio waves and television signals

antena
dispositivo que recibe ondas de radio y señales de televisión

Arctic
being from an icy climate, such as the north pole

ártico
que pertenece a un clima helado, como el del polo norte

--- B ---

behavior
all of the actions and reactions of an animal or a person (related word: behave)

conducta
todas las acciones y reacciones de un animal o una persona (palabra relacionada: comportarse)

brain
the main control center in an animal body; part of the central nervous system

cerebro
principal centro de control en el cuerpo de un animal; parte del sistema nervioso central

--- C ---

camouflage
the coloring or patterns on an animal's body that allow it to blend in with its environment

camuflaje
color o patrones del cuerpo de un animal que le permiten confundirse con su medio ambiente

canyon
a deep valley carved by flowing water

cañón
valle profundo labrado por el flujo del agua

chemical energy
energy that can be changed into motion and heat

energía química
energía que está almacenada en las cadenas entre átomos

chemical weathering
changes to rocks and minerals on Earth's surface that are caused by chemical reactions

meteorización química
cambios en las rocas y minerales de la superficie de la Tierra causados por reacciones químicas

code
a way to communicate by sending messages using dots and dashes

código
forma de comunicarse enviando mensajes con puntos y rayas

collision
the moment where two objects hit or make contact in a forceful way

colisión
el momento en el que dos objetos chocan o hacen contacto de forma contundente

Glossary | R21

conduction
when energy moves directly from one object to another

conducción
cuando la energía pasa en forma directa de un objeto a otro

conservation of energy
energy can not be created or destroyed, it can only be changed from one form to another, such as when electrical energy is changed into heat energy

conservación de la energía
la energía no se puede crear o destruir; solo se puede cambiar de una forma a otra, como cuando la energía eléctrica cambia a energía térmica

conserve
to protect something, or prevent the wasteful overuse of a resource

conservar
proteger algo o evitar el uso excesivo e ineficiente de un recurso

convert (v)
to change forms

convertir (v)
cambiar de forma

D

delta
a fan-shaped mass of mud and other sediment that forms where a river enters a large body of water

delta
masa de barro y otros sedimentos parecida a un abanico, que se forma donde un río ingresa a un gran cuerpo de agua

deposition
laying sediment back down after erosion moves it around

sedimentación
volver a depositar sedimentos una vez que la erosión los arrastra

digestive system
the body system that breaks down food into tiny pieces so that the body's cells can use it for energy

sistema digestivo
sistema del cuerpo que descompone alimentos en pequeños trozos para que las células del cuerpo puedan usarlos para obtener energía.

digital
a signal that is not continuous and is made up of tiny separate pieces

digital
una señal que no es continua y está compuesta por diminutas partes separadas

Glossary

disease
a condition that disrupts processes in the body and usually causes an illness

enfermedad
afección que perturba los procesos del cuerpo

dune
a hill of sand created by the wind

duna
colina de arena creada por el viento

---- E ----

ear
organ for hearing

oído
órgano para oir

Earth
the third planet from the sun; the planet on which we live (related words: earthly; earth – meaning soil or dirt)

Tierra
tercer planeta desde el Sol; planeta en el cual vivimos (palabras relacionadas: terrenal; tierra en el sentido de suelo o suciedad)

earthquake
a sudden shaking of the ground caused by the movement of rock underground

terremoto
repentina sacudida de la tierra causada por el movimiento de rocas subterráneas

ecosystem
all the living and nonliving things in an area that interact with each other

ecosistema
todos los seres vivos y objetos sin vida de un área, que se interrelacionan entre sí

electromagnetic spectrum
the full range of frequencies of electromagnetic waves

espectro electromagnético
rango completo de frecuencias de las ondas electromagnéticas

energy
the ability to do work or cause change; the ability to move an object some distance

energía
capacidad de hacer un trabajo o producir un cambio; capacidad de mover un objeto a cierta distancia

energy source
where a form of energy begins

fuente de energía
origen de una forma de energía

Glossary | R25

energy transfer
the transfer of energy from one organism to another through a food chain or web; or the transfer of energy from one object to another, such as heat energy

transferencia de energía
transmisión de energía de un organismo a otro a través de una cadena o red alimentaria; o transmisión de energía de un objeto a otro, como por ejemplo la energía calórica

engineer
Engineers have special skills. They design things that help solve problems.

ingeniero
Los ingenieros poseen habilidades especiales. Diseñan cosas que ayudan a resolver problemas.

environment
all the living and nonliving things that surround an organism

medio ambiente
todos los seres vivos y objetos sin vida que rodean a un organismo

erosion
the removal of weathered rock material. After rocks have been broken down, the small particles are transported to other locations by wind, water, ice, and gravity

erosión
eliminación de material rocoso desgastado. Después de descomponerse las rocas, el viento, el agua, el hielo y la gravedad transportan las partículas pequeñas a otros lugares

erupt
the action of lava coming out of a hole or crack in Earth's surface; the sudden release of hot gasses or lava built up inside a volcano (related word: eruption)

erupción
acción de la lava que sale de un agujero o cráter de la superficie de la Tierra; repentina liberación de gases calientes o lava que se acumulan en el interior de un volcán (palabra relacionada: erupción)

extinct
describes a species of animals that once lived on Earth but which no longer exists (related word: extinction)

extinto
palabra que hace referencia a una especie de animales que habitaba antiguamente la Tierra, pero que ya no existe (palabra relacionada: extinción)

— F —

fault
a fracture, or a break, in the Earth's crust (related word: faulting)

falla
fractura, o quiebre, en la corteza de la Tierra (palabra relacionada: fallas)

Glossary | R27

feature
things that describe what something looks like

rasgo
cosas que describen cómo se ve algo

---

force
a pull or push that is applied to an object

fuerza
acción de atraer o empujar que se aplica a un objeto

---

forecast
(v) to analyze weather data and make an educated guess about weather in the future; (n) a prediction about what the weather will be like in the future based on weather data

pronosticar / pronóstico
(v) analizar los datos del tiempo y hacer una conjetura informada sobre el tiempo en el futuro; (s) predicción sobre cómo será el tiempo en el futuro con base en datos

---

fossil fuels
fuels that come from very old life forms that decomposed over a long period of time, like coal, oil, and natural gas

combustibles fósiles
combustibles que provienen de formas de vida muy antiguas que se descompusieron en el transcurso de un período de tiempo largo, como el carbón, el petróleo y el gas natural

friction
a force that slows down or stops motion

fricción
fuerza que desacelera o detiene el movimiento

fuel
any material that can be used for energy

combustible
todo material que puede usarse para producir energía

—— G ——

generate
to produce by turning a form of energy into electricity

generar
producir convirtiendo una forma de energía en electricidad

geothermal
heat found deep within Earth

geotérmica
calor que se encuentra en la profundidad de la Tierra

glacier
a large sheet of ice or snow that moves slowly over Earth's surface

glaciar
gran capa de hielo o nieve que se mueve lentamente sobre la superficie de la Tierra

gravitational potential energy
energy stored in an object based on its height and mass

energía potencial gravitacional
energía almacenada debida a la ubicación en un campo gravitacional

gravity
the force that pulls an object toward the center of Earth (related word: gravitational)

gravedad
fuerza que empuja a un objeto hacia el centro de la Tierra (palabra relacionada: gravitacional)

— H —

heart
the muscular organ of an animal that pumps blood throughout the body

corazón
órgano muscular de un animal que bombea sangre a través del cuerpo

heat
the transfer of thermal energy

calor
transferencia de energía térmica

hibernate
to reduce body movement during the winter in an effort to conserve energy (related word: hibernation)

hibernar
reducir el movimiento del cuerpo durante el invierno con la finalidad de conservar la energía (palabra relacionada: hibernación)

--- I ---

information
facts or data about something; the arrangement or sequence of facts or data

información
hechos o datos sobre algo; la organización o secuencia de hechos o datos

--- K ---

kinetic energy
the energy an object has because of its motion

energía cinética
energía que posee un objeto a causa de su movimiento

Glossary | R31

L

landform
a large natural structure on Earth's surface, such as a mountain, a plain, or a valley

accidente geográfico
estructura natural grande que se encuentra en la superficie de la Tierra, como una montaña, una llanura o un valle

lava
molten rock that comes through holes or cracks in Earth's crust that may be a mixture of liquid and gas but will turn into solid rock once cooled

lava
roca fundida que sale por orificios o grietas en la corteza terrestre, y que puede ser una mezcla de líquido y gas pero se convierte en roca sólida al enfriarse

light
a form of energy that moves in waves and particles and can be seen

luz
forma de energía que se desplaza en ondas y partículas y que puede verse

M

magma
melted rock located beneath Earth's surface

magma
roca fundida que se encuentra debajo de la superficie de la Tierra

magnetic field
a region in space near a magnet or electric current in which magnetic forces can be detected

campo magnético
región en el espacio cerca de un imán o de una corriente eléctrica, donde pueden detectarse fuerzas magnéticas

map
a flat model of an area

mapa
modelo plano de un área

mass
the amount of matter in an object

masa
cantidad de materia que hay en un objeto

matter
material that has mass and takes up some amount of space

materia
material que tiene masa y ocupa cierta cantidad de espacio

meander
winding or indirect movement or course

meandro
movimiento o curso serpenteante o indirecto

Glossary | R33

migration
the movement of a group of organisms from one place to another, usually due to a change in seasons

migración
desplazamiento de un grupo de organismos de un lugar a otro, generalmente debido a un cambio de estaciones

model
a drawing, object, or idea that represents a real event, object, or process

modelo
dibujo, objeto o idea que representa un suceso, objeto, o proceso real

motion
when something moves from one place to another (related words: move, movement)

movimiento
cuando algo pasa de un lugar a otro (palabra relacionada: mover, desplazamiento)

mountain
an area of land that forms a peak at a high elevation (related term: mountain range)

montaña
área de tierra que forma un pico a una elevación alta (palabra relacionada: cadena montañosa)

N

nerve
a cell of the nervous system that carries signals to the body from the brain, and from the body to the brain and/or spinal cord

nervio
célula del sistema nervioso que lleva señales al cuerpo desde el cerebro, y desde el cuerpo al cerebro y/o médula espinal

nonrenewable
once it is used, it cannot be made or reused again

no renovable
una vez usado, no puede rehacerse o reutlizarse

nonrenewable resource
a natural resource of which a finite amount exists, or one that cannot be replaced with currently available technologies

recurso no renovable
recurso natural del cual existe una cantidad finita, o que no puede remplazarse con las tecnologías actualmente disponibles

nuclear energy
the energy released when the nucleus of an atom is split apart or combined with another nucleus

energía nuclear
energía liberada cuando el núcleo de un átomo se divide o combina con otro núcleo

O

ocean
a large body of salt water that covers most of Earth

océano
gran cuerpo de agua salada que cubre la mayor parte de la Tierra

opaque
describes an object that light cannot travel through

opaco
describe un objeto que la luz no puede atravesar

organ
a group of tissues that performs a complex function in a body

órgano
conjunto de tejidos que realizan una función compleja en el cuerpo

organism
any individual living thing

organismo
todo ser vivo individual

P

photosynthesis
the process in which plants and some other organisms use the energy in sunlight to make food

fotosíntesis
proceso por el cual las plantas y algunos otros organismos usan la energía del Sol para producir alimentos

pollute
to put harmful materials into the air, water, or soil (related words: pollution, pollutant)

contaminar
poner materiales perjudiciales en el aire, agua o suelo (palabras relacionadas: contaminación, contaminante)

pollution
when harmful materials have been put into the air, water, or soil (related word: pollute)

contaminación
cuando se introducen materiales perjudiciales en el aire, el agua o el suelo (palabra relacionada: contaminar)

potential energy
the amount of energy that is stored in an object; energy that an object has because of its position relative to other objects

energía potencial
cantidad de energía almacenada en un objeto; energía que tiene un objeto debido a su posición relativa con otros objetos

predator
an animal that hunts and eats another animal

depredador
animal que caza y come a otro animal

Glossary | R37

predict
to guess what will happen in the future (related word: prediction)

predecir
adivinar qué sucederá en el futuro (palabra relacionada: predicción)

prey
an animal that is hunted and eaten by another animal

presa
animal que es cazado y comido por otro

pupil
the black circle at the center of an iris that controls how much light enters the eye

pupila
círculo negro en el centro del iris que controla cuánta luz entra al ojo

— R —

radiant energy
energy that does not need matter to travel; light

energía radiante
energía que no necesita de la materia para viajar; luz

radiation
electromagnetic energy (related word: radiate)

radiación
energía electromagnética (palabra relacionada: irradiar)

receptor
nerves located in different parts of the body that are especially adapted to receive information from the environment

receptor
nervios ubicados en diferentes partes del cuerpo que están especialmente adaptados para recibir información del medio ambiente

reflect
light bouncing off a surface (related word: reflection)

reflejar
rebotar la luz sobre una superficie (palabra relacionada: reflexión)

reflex
an automatic response

reflejo
respuesta automática

refract
to bend light as it passes through a material (related word: refraction)

refractar
torcer luz cuando pasa a través de un material (palabra relacionada: refracción)

remote (adj)
to be operated from a distance

remoto (adj)
que se opera a distancia

renewable
to reuse or make new again

renovable
reutilizar o volver a hacer de nuevo

renewable resource
a natural resource that can be replaced

recurso renovable
recurso natural que puede reemplazarse

reproduce
to make more of a species; to have offspring (related word: reproduction)

reproducir
engendrar más individuos de una especie; tener descendencia (palabra relacionada: reproducción)

resistance
when materials do not let energy transfer through them

resistencia
cuando los materiales no permiten la transferencia de energía a través de ellos

resource
a naturally occurring material in or on Earth's crust or atmosphere of potential use to humans

recurso
material que se origina de forma natural en o sobre la corteza o la atmósfera de la Tierra, que es de uso potencial para los seres humanos

rotate
turning around on an axis; spinning (related word: rotation)

rotar
girar sobre un eje; dar vueltas (palabra relacionada: rotación)

S

satellite
a natural or artificial object that revolves around another object in space

satélite
objeto natural o artificial que gira alrededor de otro objeto en el espacio

sediment
solid material, moved by wind and water, that settles on the surface of land or the bottom of a body of water

sedimento
material sólido que el viento o el agua transportan y que se asienta en la superficie de la tierra o en el fondo de un cuerpo de agua

seismic
having to do with earthquakes or earth vibrations

sísmico
relativo a los terremotos o a las vibraciones de la Tierra

seismic wave
waves of energy that travel through the Earth

onda sísmica
ondas de energía que se desplazan a través del interior de la Tierra

senses
taste, touch, sight, smell, and hearing (related word: sensory)

sentidos
gusto, tacto, visión, olfato y oído (palabra relacionada: sensorial)

skin
an organ that covers and protects the bodies of many animals

piel
órgano que cubre y protege el cuerpo de muchos animales

soil
the outer layer of Earth's crust in which plants can grow; made of bits of dead plant and animal material as well as bits of rocks and minerals

suelo
capa externa de la corteza de la Tierra en donde crecen las plantas; formada por pedazos de plantas y animales muertos, así como por pedazos de rocas y de minerales

solar energy
energy that comes from the sun

energía solar
energía que proviene del Sol

sound
anything you can hear that travels by making vibrations in air, water, and solids

sonido
todo lo que se puede oír, que se desplaza produciendo vibraciones en el aire, el agua y los objetos sólidos

sound wave
a sound vibration as it is passing through a material: Most sound waves spread out in every direction from their source.

onda sonora
vibración que produce el sonido cuando atraviesa un material: la mayoría se dispersa desde la fuente en todas direcciones.

speed
the measurement of how fast an object is moving

rapidez
medida de la tasa a la que se desplaza un objeto

stimulus
things in the environment that cause us to react or have a physical response

estímulo
algo en el medio ambiente que nos hace reaccionar o tener una respuesta física

stomach

a muscular organ in the body where chemical and mechanical digestion take place

estómago

órgano muscular del cuerpo donde tiene lugar la digestión química y mecánica

sun

any star around which planets revolve

sol

toda estrella alrededor de la cual giran los planetas

survive

to continue living or existing: an organism survives until it dies; a species survives until it becomes extinct (related word: survival)

sobrevivir

continuar viviendo o existiendo: un organismo sobrevive hasta que muere; una especie sobrevive hasta que se extingue (palabra relacionada: supervivencia)

system

a group of related objects that work together to perform a function

sistema

grupo de objetos relacionados que funcionan juntos para realizar una función

T

tectonic plate
one of several huge pieces of Earth's crust

placa tectónica
una de las muchas partes enormes de la corteza terrestre

thermal energy
energy in the form of heat

energía térmica
energía en forma de calor

tongue
an organ in the mouth that helps in eating and speaking

lengua
órgano de la boca que ayuda a comer y hablar

topographic map
a map that shows the size and location of an area's features such as vegetation, roads, and buildings

mapa topográfico
mapa que muestra el tamaño y la ubicación de características de un área, como la vegetación, las carreteras y los edificios

trait
a characteristic or property of an organism

rasgo
característica o propiedad de un organismo

Glossary | R45

transparent
describes materials through which light can travel; materials that can be seen through

transparente
describe materiales a través de los cuales puede desplazarse la luz; materiales a través de los cuales se puede ver

tsunami
a giant ocean wave (related word: tidal wave)

tsunami
ola gigante en el océano (palabra relacionada: maremoto)

--- V ---

valley
a low area of land between two higher areas, often formed by water

valle
área baja de tierra entre dos áreas más altas, generalmente formada por el agua

volcano
an opening in Earth's surface through which magma and gases or only gases erupt (related word: volcanic)

volcán
abertura en la superficie de la Tierra a través de la cual surgen magma y gases, o solo gases, que hacen erupción (palabra relacionada: volcánico)

W

water
a compound made of hydrogen and oxygen; can be in either a liquid, ice, or vapor form and has no taste or smell

agua
compuesto formado por hidrógeno y oxígeno; puede estar en forma de líquido, hielo o vapor y no tiene sabor ni olor

wave
a disturbance caused by a vibration; waves travel away from the source that makes them

onda
perturbación causada por una vibración que se aleja de la fuente que la origina

wavelength
the distance between one peak and the next on a wave

longitud de onda
distancia entre un pico y otro en una onda

weathering
the physical or chemical breakdown of rocks and minerals into smaller pieces or aqueous solutions on Earth's surface

meteorización
desintegración física o química de rocas y minerales en partes más pequeñas o en soluciones acuosas en la superficie de la Tierra

work
a force applied to an object over a distance

trabajo
fuerza aplicada a un objeto a lo largo de una distancia

Index

A

Air 17, 176
Alaska 179
Analyze Like a Scientist 17–19, 29–30, 34–35, 42–45, 66–67, 76–77, 78–79, 87–93, 100–103, 112–113, 118–119, 126–128, 129–130, 142–148, 157–160, 164–165, 172–173, 176–178, 181–182, 190–192
Ask Questions Like a Scientist 10–11, 52–53, 110–111, 154–155

B

Badlands 27

C

Can You Explain? 8, 39–41, 50, 97–99, 108, 141–143, 152, 187–189
Canyon 66-68, 196
Chemical Weathering 17, 22, 25
Chesapeake Bay 138
Continental Divide 130
Contour line 119, 134
Crust 164, 172–173

D

Delta 34, 70
Deposition 34–35, 42, 70
Divide 130
Dune 35, 79–80

E

Earth 190
Earthquake 182
Elevation 116, 134–136
Erosion 29–31, 34, 42, 76–79
Erupt 158, 173, 175–176, 179, 181–183, 191
    Alaskan volcanoes 179
    Mt. Vesuvius 158
    studying eruptions 173, 175–176, 181–183, 191
Estuary 126
Evaluate Like a Scientist 15–19, 36–37, 46, 60–64, 70–71, 94–95, 104, 115–116, 134–137, 149, 160–161, 179–180, 184, 194–195

F

Fossil 87, 89–91

G

Geology 42
Glacier 29–31, 35, 76
Grand Canyon 4, 196

H

Hands-On Activities 22–24, 31–33, 54–57, 72–75, 80–82, 83–86, 122–125, 131–133, 168–171
Heat 17

I

Investigate Like a Scientist 22–24, 31–33, 54–57, 72–75, 80–82, 83–86, 122–125, 131–133, 168–171

L

Landforms (landscape)
    changes in 76–79, 94–95, 100–101

mapping 72, 115–116, 118
types of 67, 76–79, 94–95, 112, 129–130
Lava 91, 172–173, 176–177

M

Magma 173, 182
Maps
elevation 134–138
using 118–120, 126–127, 144–146
Mechanical weathering 18, 25
Mount Vesuvius 154, 158, 186
Mountain 112, 118

O

Observe Like a Scientist 12–14, 16, 20–21, 25–26, 27–28, 58–59, 64–65, 68–69, 114, 120–121, 138–139, 156, 162–163, 166–167, 174, 175, 183
Ocean 112–114, 144–145, 164

P

Pangaea 89
Photography 100–102
Pompeii 157–158

R

Rain 67, 130
Record Evidence Like a Scientist 38–41, 96–99, 140–143, 186
Rivers 131
Rock 83, 87–93

S

Sailing 146
Sand 80
Sandcastle 10, 12, 38
Satellite 126
Sediment
from deposition 34–35, 83
from erosion 29–30, 67, 78, 87
Seismometer 182
Soil 29
Solve Problems Like a Scientist 4–5, 196–199
Sonar 146, 148
Streams
causing erosion and deposition 67, 70, 89–90
creating other landforms 67, 72, 129–130
Stream table 72

T

Technology 103
Tectonic plate 164
Tiltmeter 181–182
Topographic map 118, 122, 162

U

Unit Project 4–5, 196–199

V

Valley 70–72, 76, 129
Volcanoes 89, 91, 158, 160–164, 166, 172–174, 176–179, 180–181, 183–184, 190–192, 194

Index | R51

Index

    eruptions 158, 176–179
    forming 89, 91, 160–164, 166
    properties of 172–174
    studying 190–192
Volcanologist 192

W

Water
    erosion 17, 29–30, 76, 89
    in land formations 67, 70, 89, 129
    on maps 118, 127, 144

Watersheds 129–131, 138
Weathering 16–18, 22, 29, 42
Wetland 70
Wind 78, 80

Z

Zion 87–88